I-35W Bridge Collapse and Response
Minneapolis, Minnesota
August 1, 2007

Reported by: Hollis Stambaugh
Harold Cohen

This is Report 166 of Investigation and Analysis of Major Fire Incidents and USFA's Technical Report Series Project conducted by TriData, a Division of System Planning Corporation under Contract (GS-10-F0350M/HSFEEM-05-A-0363) to the DHS/U.S. Fire Administration (USFA), and is available from the USFA Web site at http://www.usfa.dhs.gov

 **FEMA**

Department of Homeland Security
United States Fire Administration
National Fire Programs Division

U.S. Fire Administration Fire Investigations Program

The U.S. Fire Administration (USFA) develops reports on selected major fires throughout the country. The fires usually involve multiple deaths or a large loss of property, but the primary criterion for deciding to write a report is whether it will result in significant "lessons learned." In some cases these lessons bring to light new knowledge about fire—the effect of building construction or contents, human behavior in fire, etc. In other cases, the lessons are not new, but are serious enough to highlight once again because of another fire tragedy. In some cases, special reports are developed to discuss events, drills, or new technologies or tactics that are of interest to the fire service.

The reports are sent to fire magazines and are distributed at national and regional fire meetings. The reports are available on request from USFA. Announcements of their availability are published widely in fire journals and newsletters.

This body of work provides detailed information on the nature of the fire problem for policymakers who must decide on allocations of resources between fire and other pressing problems, and within the fire service to improve codes and code enforcement, training, public fire education, building technology, and other related areas.

The USFA, which has no regulatory authority, sends an experienced fire investigator into a community after a major incident only after having conferred with the local fire authorities to ensure that USFA's assistance and presence would be supportive and would in no way interfere with any review of the incident they are themselves conducting. The intent is not to arrive during the event or even immediately after, but rather after the dust settles, so that a complete and objective review of all the important aspects of the incident can be made. Local authorities review USFA's report while it is in draft form. The USFA investigator or team is available to local authorities should they wish to request technical assistance for their own investigation.

For additional copies of this report write to the U.S. Fire Administration, 16825 South Seton Avenue, Emmitsburg, Maryland 21727 or via http://www.usfa.dhs.gov

U.S. Fire Administration

Mission Statement

As an entity of the Federal Emergency Management Agency (FEMA), the mission of the U.S. Fire Administration (USFA) is to reduce life and economic losses due to fire and related emergencies, through leadership, advocacy, coordination, and support. We serve the Nation independently, in coordination with other Federal agencies, and in partnership with fire protection and emergency service communities. With a commitment to excellence, we provide public education, training, technology, and data initiatives.

ACKNOWLEDGMENTS

The information for this report was derived in large part from individuals who were interviewed in the Twin Cities area on site and via videoconference over a period of 3 days. The U.S. Department of Homeland Security (DHS) provided videoconferencing capabilities for those interviews and linked into the system made available by the State of Minnesota. Additional interviews were conducted by telephone. The names of the people who were interviewed are provided in Table 1 below.

We also reviewed hundreds of pages of documentation on the response and recovery activities, including situation reports from the Minneapolis and State of Minnesota Emergency Operations Centers (EOCs), internal after-action analyses, maps, articles, and presentations that provided additional insight and details. The authors are grateful for the tremendous support offered by all, especially key Minneapolis emergency preparedness personnel, Rocco Forté and Bill Anderson, and Robert Berg from Minnesota Homeland Security and Emergency Management.

Table 1: Individuals Interviewed for Report

City of Minneapolis	
Allen, Rob	Police Department
Anderson, Bill	Emergency Preparedness
Blixt, Pam, Manager	Health and Family Support Preparedness
Bundt, Jonathan	Police Department, Consultant/Hospital Compact Group
Dejung, John	9-1-1/3-1-1
Dietrich, Sara	Communications
Dressler, Lisa	Emergency Preparedness
Eicklenberg, Laura	Emergency Preparedness
Forté, Rocco, Director	Emergency Preparedness
Fruetel, John	Fire Department
Hermanson, Stacy	Emergency Preparedness
Johnston, Heather	Finance Department
Kennedy, Mike	Public Works
Laible, Matt	Communications
Martin, Mike	Police Department
Musicant, Gretchen	Health Department
Rollwagen, Kristi	Emergency Preparedness
Sobania, Don	Public Works

continued on next page

Stewart, Dr. Jeffrey, Chaplain	Police Department
Velasco-Thompson, Ellen	Risk Management and Crisis
Wagenpfeil, Otto	Police Department

Hennepin County

Chandler, Bill	Sheriff's Office
Geisehardt, Roberta	Medical Examiner
Turnbull, Tim	Emergency Preparedness
Van Buren, Martin	Emergency Medical Services
Ward, Tom	Emergency Medical Services

State Homeland Security and Emergency Management

Berg, Robert M.
Berrisford, David
Eide, Kris
Hendrickson, Gary
Ketterhagen, Kim
Lokken, Garry

U.S. Coast Guard, Upper Mississippi River Sector

Epperson, Todd
Richey, Sharon, Captain

TABLE OF CONTENTS

CHAPTER I. OVERVIEW OF THE EVENT AND MINNEAPOLIS'S PREPAREDNESS 1

CHAPTER II. FIRE AND EMERGENCY MEDICAL SERVICES 5

    Initial Fire and Rescue Response 6

    Emergency Medical Services 9

CHAPTER III. LAW ENFORCEMENT AND THE FAMILY ASSISTANCE CENTER 17

CHAPTER IV. RECOVERY OPERATIONS AND THE MEDICAL EXAMINER'S OFFICE 22

    Medical Examiner's Office 26

    Fatalities 27

    Antemortem and Postmortem Data 27

    Respect and Sensitivity to Victims' Families 28

CHAPTER V. MINNEAPOLIS EMERGENCY MANAGEMENT SYSTEM 29

    Minneapolis Emergency Operations Center 29

    Roles and Relationships 30

    Emergency Operations Center Oversight 31

    Safety 32

    Public Information Officer 32

    Liaison 32

    Operations 33

    Planning 33

    Logistics and Public Works 35

    9-1-1 Center 36

    Finance and Administration 38

    Minnesota Security and Emergency Management 39

CHAPTER VI. HAZMAT AND ENVIRONMENTAL MONITORING 41

CHAPTER VII. LESSONS LEARNED AND BEST PRACTICES 43

    Problem Areas and Lessons Learned 43

    Best Practices: Notable Successes in the Response to Bridge Collapse . . . 44

CHAPTER I. OVERVIEW OF THE EVENT AND MINNEAPOLIS'S PREPAREDNESS

In 1967, the Interstate 35W Mississippi River Bridge in Minneapolis opened to traffic. The bridge was 1,907 feet long, had 14 spans, and by 2007 carried a daily average of 140,000 total vehicles north and south over four lanes between University Avenue and Washington Avenue. The vehicle count made it one of the busiest bridges in the country over the Mississippi River, and one of three principal arteries into downtown Minneapolis, a city with one of the highest population densities in the Midwest.

Just after 6 p.m. on the evening of August 1, 2007, the 40-year old bridge collapsed into the river and its banks without warning, killing 13 and injuring 121 others. At the time, there were approximately 120 vehicles, carrying 160 people, on the bridge. The impact of the fall broke the span into multiple planes of broken steel and crushed concrete—cars, buses, and trucks all resting precariously along guardrails or suddenly unprotected edges, crashed into other vehicles, partially embedded in the muddy river bank, or dropped precipitously into the river (Figure 1).

Figure 1. Scene of the Collapse.

The most urgent task was rescuing people from the water and from their vehicles, conducting triage on the injured, and providing transport to area hospitals. Several vehicles were on fire, making fire-fighting operations a parallel priority. Local and State staff and officials from fire, law enforcement, emergency management, and public works received immediate alerts and, having trained together in classroom settings and through field exercises, knew what to do and with whom they needed to coordinate their response. Years of investing time and money into identifying gaps in the city's disaster preparedness capabilities; acquiring radios for an interagency, linked 800 MHz system; and participating in training on the National Incident Management System (NIMS) and on the organizational basis for that system (the Incident Command System (ICS) and Unified Command) paid off substantially during response and recovery operations.

The bridge collapse tested the area's ability to handle a complicated mix of tactical and strategic problems effectively. People and vehicles fell into the water, onto the river banks, and onto the multiple surfaces of the broken bridge (Figure 2). It was a dangerous multilevel accident site, located in the river gorge, that had the ever-present potential for secondary collapse. The fires on the deck; the presence of certain hazardous materials and the prospect of others; steep banks; and collapse debris complicated access, among other difficulties (Figure 3). The bridge was owned by the Federal government, and was operated and maintained by the State of Minnesota. After the collapse, the bridge was lodged in the river where, under Minnesota statute, the Hennepin County Sheriff's Office Water Patrol has jurisdiction; and along the river banks, which are under the jurisdiction of the City of Minneapolis.

Figure 2. One Section of Collapsed Bridge.

The City of Minneapolis and the Hennepin County Sheriff's Office were assisted by a multitude of mutual-aid resources from adjacent counties and cities, and by State and Federal agencies. Federal assets soon at the scene included the U.S. Army Corps of Engineers, the U.S. Coast Guard, and the U.S. Navy.

The excellent working relationships that had been developed through joint interagency training, planning, and previous emergency incidents was one of the primary reasons that response and recovery operations went as smoothly as they did. As one leader commented "We didn't view it as a Minneapolis incident; it was a city/county/State incident."

Figure 3. Another View of the Bridge Collapse.

The local response to the bridge disaster—and the coordination with metro, State, and Federal partners—demonstrated the extraordinary value of comprehensive disaster planning and training. The City of Minneapolis was as well prepared as any local jurisdiction could be to handle a major incident. The city's ability to respond had evolved over several years of investing heavily and widely in all the elements that make a crucial difference when disaster strikes. Their investment covered widespread training on the NIMS that extended beyond city department heads and into all employee levels. Over half of the city's 4,000 employees have received NIMS training.

In 2002, Minneapolis elected officials and key staff took a hard look at its state of preparedness following an intensive, 1-week Federal Emergency Management Agency (FEMA) offsite training course at Mt. Weather in Virginia, and conducted a risk assessment that identified areas where improvements were needed. The city wasted no time in resolving the gaps, aggressively pursuing Federal grant dollars, e.g., the Urban Area Security Initiative, and general fund dollars to pay for radio and communications upgrades, equipment, and training that together elevated its level of preparedness.

Minneapolis is part of what now is identified as the Twin Cities Urban Area (TCUA) for the purpose of Federal homeland security grant programs. The TCUA includes all the jurisdictions that provided mutual aid in some capacity to Minneapolis and Hennepin County when the bridge disaster occurred. Ramsey, Hennepin, and Dakota Counties along with the City of Saint Paul and Minneapolis make up the TCUA. Within this metropolitan region of three million people are two bomb squads, two chemical assessment teams, a heavy and medium collapsed-structure rescue team, a type three and type four all-hazard incident management team and the State's only statewide hazardous materials emergency response team.

Federal grants from the Department of Homeland Security (DHS) to the TCUA have benefited both Minneapolis and the region as a whole. Those grants have supported the development of the special capabilities teams mentioned above, as well as training, planning, and resources for transit and port security, a metropolitan medical response team, Continuity of Operations (COOP) and Continuity of Government (COG) planning, incident command, NIMS training, and more.

As was mentioned before, in March of 2002, 80 top officials from Minneapolis, Hennepin County, and the State of Minnesota attended a 4-day Integrated Emergency Management course. The purpose of the course was to test the city's Emergency Operations Plan (EOP) and identify weaknesses. Using multiple scenarios, the EOP that existed at the time was put through its paces and exercised by the entire group of city stakeholders. Following this experience, the city set up work teams under police, fire, public works, and communications to address the shortcomings they identified during training. Following are some of the specific actions they took relative to some weak areas:

1. **Communications**—Minneapolis earmarked $20 million to purchase new 800 MHz radios. Not surprisingly, Minneapolis was ranked as one of six leading cities in tactical interoperability communications, according to a recent Federal report.

2. **Emergency Dispatch**—The city spent $5.2 million on a state-of-the-art computer-aided dispatch (CAD) system that has the capability to map the location of all emergency response vehicles equipped with Global Positioning Systems (GPS).

3. **Special Operations Teams**—The exercises from training led to the realization that Minneapolis needed specialized teams with technical skills that would be used in various disaster situations. The city created three special response teams: the fire department's hazardous materials and collapse structure teams, and the police department's bomb squad, at a cost of $8 million.

4. **Pharmaceutical Stockpile**—Minneapolis now has a comprehensive plan for storing and distributing pharmaceuticals in the event of widespread disease.

The city made progress on infrastructure protection—hardening soft targets—and training. All city employees were given the opportunity for training on the ICS and how that structure functions in the field and in the Emergency Operations Center (EOC). Employees' awareness of the main elements of ICS contributed to smoother response and recovery than would otherwise have been the case.

When key personnel from the primary response agencies were asked to what they attributed their excellent response, without exception they answered, "relationships." Those relationships were developed as a result of all the planning, training, and exercises that multiple agencies and levels of government shared in recent years. Responders knew whom to call for what resources. They knew to work through the established chain of command. They knew each other's names and faces and had built a level of trust that made it possible to move quickly through channels and procedures.

The main task that remains is to locate adequate space and fully equip a new EOC so that all the players on the emergency management support team can operate physically from the same location.

This report documents how Minneapolis used the many resources at its disposal, managed firefighting and rescue operations, controlled perimeters and maintained security, recovered fatalities from the river, and handled hazardous materials concerns and safety overall. The report also discusses the support provided to the families of the dead and describes how the emergency management system worked at the city's EOC. Several best practices are described at the end of the report. The U.S. Fire Administration (USFA) also believes there are important lessons to be learned from some problems that occurred. These are discussed in the final chapter as well.

CHAPTER II. FIRE AND EMERGENCY MEDICAL SERVICES

This chapter discusses the onscene fire, rescue, and emergency medical services (EMS) operations and how the Incident Command System (ICS) was used. While the incident was complex, with recovery and debris management activities that extended for weeks, the rescue phase was completed in a relatively short time. Emergency responders were able to assess the situation quickly, determine priorities, and perform needed rescue and emergency care.

All responders had received prior training in the MnNIMS, Minnesota's version of the National Incident Management System (NIMS). Operational leaders, administrators, and responders had completed required training according to their levels of responsibility.[1]

The biggest challenge facing emergency responders was the need to bring order to a chaotic situation. Upon their arrival, police, fire, and EMS units were faced with a large structural collapse, multiple deaths and injuries, vehicle fires, entrapments, and water rescue situations (Figure 4). Authorities were uncertain whether the disaster they were facing might be the work of domestic or foreign terrorists or the failure of infrastructure due to another cause. There also were serious concerns about the possibility of further collapse, or a secondary incident on the adjacent bridge.

Figure 4. Fire Suppression at the Bridge.

Initially, separate Incident Command Posts (ICPs) were established by the fire and police departments. EMS at first reported to the north side of the bridge collapse zone, but was in consultation with the Hennepin County Medical Center (HCMC) EMS Director who was the liaison at fire Command. Police Command and the transition to a full Unified Command are discussed later, as are the activities of the Hennepin County Sheriff's Office and the Minneapolis Emergency Management Agency.

---

[1] As of February 8, 2008, the MnNIMS program was discontinued, and Minnesota adopted the Federal NIMS program.

INITIAL FIRE AND RESCUE RESPONSE

On August 1, 2007, at 6:06 p.m., the Minneapolis Fire Department (MFD) was dispatched to the I-35W Bridge for a reported bridge collapse. Fire dispatch notified responding units that they had received several calls advising of major injuries, major structural damage, and vehicles in the water. Dispatch also received video from bridge video cameras, usually used to monitor traffic conditions, which also indicated the extent of the incident. By 6:11 p.m., MFD units arrived on scene. They began implementation of the ICS and assessment of the devastation before them. At 6:16 p.m., Engine 11 requested a second-alarm assignment, and five additional MFD units were dispatched.

By 6:18 p.m., additional engine, truck, and rescue companies arrived, including onduty battalion and deputy chiefs. Deputy Chief 2 assumed Incident Command, while Battalion Chief 1 was assigned Division 3 and Battalion Chief 3 was assigned Division 4. Fire units reported the following situations to Command:

- Total bridge collapse with cars in the water, many injuries.

- Engine 6 laying a supply line from the southwest side of the bridge.

- Rescue 9 checking damaged railroad cars for contents and potential hazardous materials release.

- Ladder 3 reporting fire on the south side of the bridge.

- One patient on the bridge is in cardiac arrest, with cardiopulmonary resuscitation (CPR) in progress.

- Rescue 1 reporting that upper and lower lock boats were being used for rescue.

At 6:25 p.m. the MFD Assistant Chief of Operations assumed Command and set up a Command Post on the 10th Avenue Bridge, adjacent to the I-35W Bridge. Minneapolis police elected to set up a separate CP closer to the I-35W Bridge, in front of the Red Cross Building.

The MFD Incident Commander (IC) determined that the 10th Avenue Bridge (Figure 5) would provide the best visibility of the entire incident, less congestion from vehicles parked closer to the site, and room to set up a Unified Command Post. From that vantage point, the fire IC could have a better picture of rescue, fire, and EMS operations.

Shortly after establishing the new CP, Command began to expand the ICS. The initial organization is shown in Figure 6.

The situation confronting each group was intense, with multiple rescues and multiple casualties reported. Most of the casualties were reported from the South Deputy Group located at the northern part of the bridge. Command was able to determine a status for each group (Table 2).

Figure 5. View of the 10th Avenue Bridge.

Figure 6. Early MFD Incident Command Chart at the 10th Avenue Bridge.

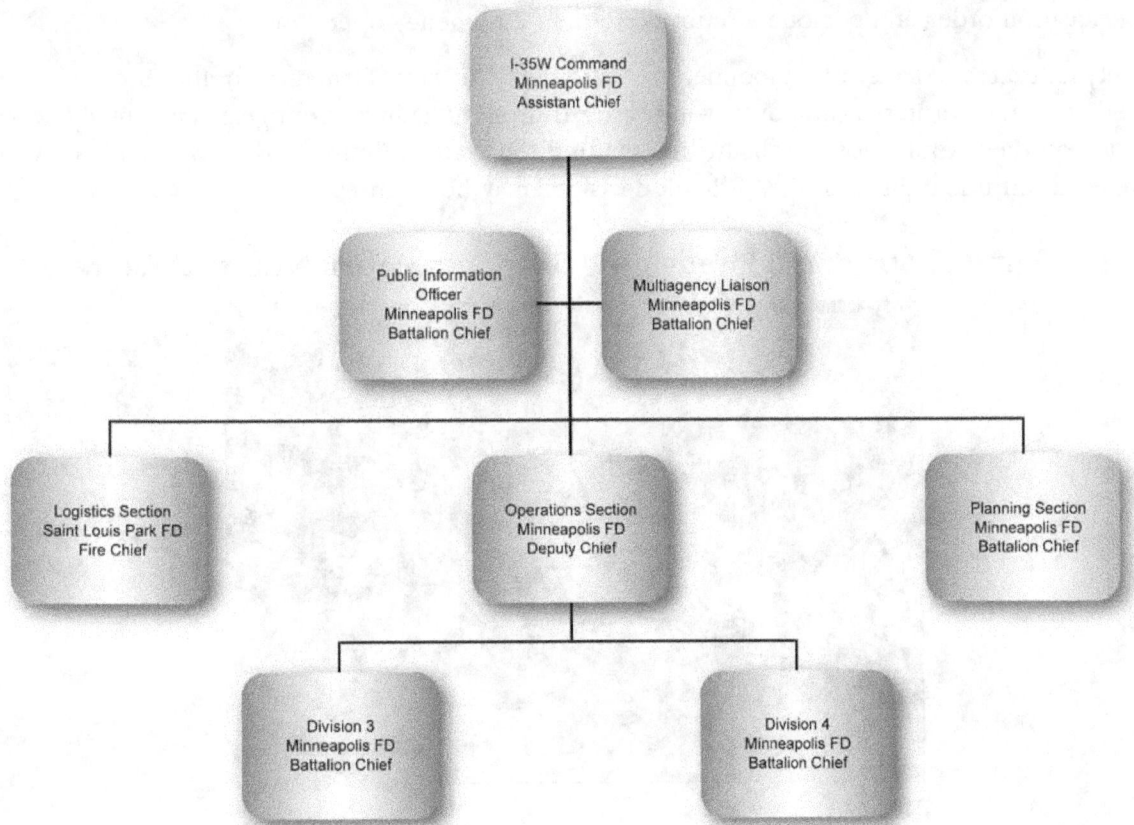

Table 2: Status Reports as of August 1—6:50 p.m.

Operational Group	Status
Division 3	At least five working vehicle fires; school bus crashed with up to 35 injured; several entrapped patients in vehicles; confined space rescue situations; bridge unstable
Division 4	Collapse with entrapment; several crush injuries
Division 3	School bus and cars involved; fires are being contained
Division 3	Many vehicles involved; primary search completed; one vehicle rolled off the bridge; possible bridge instability; structural engineer on scene
Division 3	Five boats in operation; uninjured moved by boat; divers in the water assisted with confined space rescue

By 7 p.m. there were several reports of bridge instability. The engineer working with Division 3 confirmed that bridge stability was questionable, and the fire IC then ordered all personnel to leave the bridge until a further evaluation could be made. The order was transmitted by an all-call announcement on all talk-group channels and simultaneous sounding. Firefighters immediately complied with the evacuation order. Unfortunately, other responders had difficulty leaving entrapped patients or rescue activities and did not evacuate the bridge. Non-fire responders may not have heard the radio broadcast or understood the meaning of the horns. They also may have decided to stay with patients, despite the danger. Based on feedback from firefighters, engineers, and others, the fire IC made the appropriate decision, and it did not have a negative effect on the outcome of any victim. After consulting with other public safety departments, engineers, and public works, the fire IC lifted the evacuation order, and personnel returned to their emergency operations.

A safety issue arose concerning responder protection near the river. Structural firefighting gear should not be worn near or in standing or moving water. If firefighters in turnout gear accidentally became submerged they would face a difficult time exiting the water (Figure 7). During incidents where there is an imminent threat to life, difficult decisions may be required.

Figure 7. Structural Firefighting Gear Being Worn Near the Water.

The fire IC was in close contact with the Emergency Operations Center (EOC) that was attempting to determine the support needed for the incident. Fire personnel continued to fight fires and perform extrication. While no formal Safety Officer was appointed, the IC and all Staff officers dedicated significant time to victim and responder safety. This included monitoring of bridge and debris stability, checking that personal protective equipment (PPE) was worn, ensuring the safe use of tools, and ascertaining the safety of injured and entrapped patients. Fire personnel also assisted EMS crews with triage and treatment of victims. Mutual-aid companies from a number of other jurisdictions assisted with patient care, rescue, and hazardous materials mitigation and control. By 7:55 p.m. the last live rescue victim was transported from the scene.

EMERGENCY MEDICAL SERVICES

Hennepin County Emergency Medical Services provides EMS for much of Hennepin County and most of the City of Minneapolis. The service operates as a combination municipal/hospital-based model with a communications system separate from fire and law enforcement. Nine EMS agencies responded directly to the incident or provided mutual-aid coverage for the city. An estimated 31 ambulances responded to the scene, including 12 from HCMC and 12 from mutual-aid partners Allina Medical Transportation and North Memorial Ambulance (Figure 8).

Figure 8. EMS Personnel Assist a Victim.

At 6:05 p.m. Hennepin EMS received a call via the computer–aided dispatch (CAD) from Minneapolis 9-1-1 for a possible bridge collapse. The location assigned, based on the cell–phone caller, was 500 2nd Street, SE, five blocks away from the bridge. The initial response included Medics 421, 481, 485, 488, and the onduty supervisor. The first units to arrive had obstructed views and did not realize the magnitude of the situation. Medic 485 responded across the Stone Arch Bridge, upstream from

the bridge collapse, and was the first EMS unit able to observe the entire bridge span. They diverted from the initial address and used a private road to access the river flats. Medic 485 then established the North EMS Division and provided dispatch with an incident sizeup. Crewmembers took on the roles of Triage Supervisor and Transportation Supervisor. The 485 Triage Supervisor climbed onto the debris field, reported additional casualty information, and coordinated patient rescue and extrication with fire and police.

Emergency Medical Services Branch–The EMS onduty supervisor provided a more extensive sizeup and assumed the EMS Branch Director position. The road to the river flat below the power plant was narrow and had a blind corner. Staging was located at Main Street, SE, and 6th Avenue, SE, about one-quarter mile from the river flat. Medic 485 previously had requested a second radio talk group, but the EMS Branch reversed that decision because there was insufficient Command structure set up to support a multiple talk group operation.

The EMS Branch reported that there were an estimated 20 to 30 injuries, requested more resources, and held all crews about to go off duty. While the majority of EMS crews were assigned to the bridge, service was maintained throughout Hennepin EMS's coverage area using unassigned crews and mutual aid.

EMS dispatch paged the department's management team. Unit 401, the Hennepin EMS Director, responded with the first crews and reported to the 10th Avenue Bridge as fire arrived to establish a CP while the Assistant Director stayed behind to manage department activities. Unit 401 assumed the liaison position between the EMS Branch and fire Command. From their position on the 10th Avenue Bridge EMS liaison and fire IC were able to see the extent of the bridge collapse clearly and to monitor many of the activities of the rescue operation directly. [2]

Northeast Division: Medic 482 responded to the river flat in Northwest Division. After reporting to 485 Transport Supervisor, they left their ambulance, advanced across the debris pile, and established the Northeast Division. There were many critical patients, numerous difficult extrications, and two deceased patients. On the north side of the river the best access to the collapse zone was on the downriver side in the Northwest Division. Access and egress from the Northeast Division was limited because the usual access to the area was blocked by the collapsed bridge and the alternate route (through an off-road construction zone) was not widely known by most responders.

Medic 482 established another Triage and Transportation Unit. The first critical patients were transported by a mutual-aid ambulance that had followed a fire truck onto the lower river flat. This ambulance made two round trips to HCMC. Despite the efforts of EMS Branch, Transport Supervisors, dispatch, and numerous crews, no other ambulance was successful in finding this route during the rescue phase of the response. The decision to use alternate transportation was made jointly between the Transport Supervisor, EMS Medical Director, and EMS Branch. A few pickup trucks had appeared in the downriver flat area and were put into service for this purpose. An estimated 8 to 13 patients were transported this way with first responders accompanying the patients. Ambulances successfully rendezvoused with at least one pickup and continued to transport critical patients (Figure 9).

---

[2]Van Buren, M., et al. (2007). *I-35W Bridge Collapse: After Action Report.* Unpublished Manuscript. Hennepin Emergency Medical Services.

Figure 9. EMS Transport of a Patient.

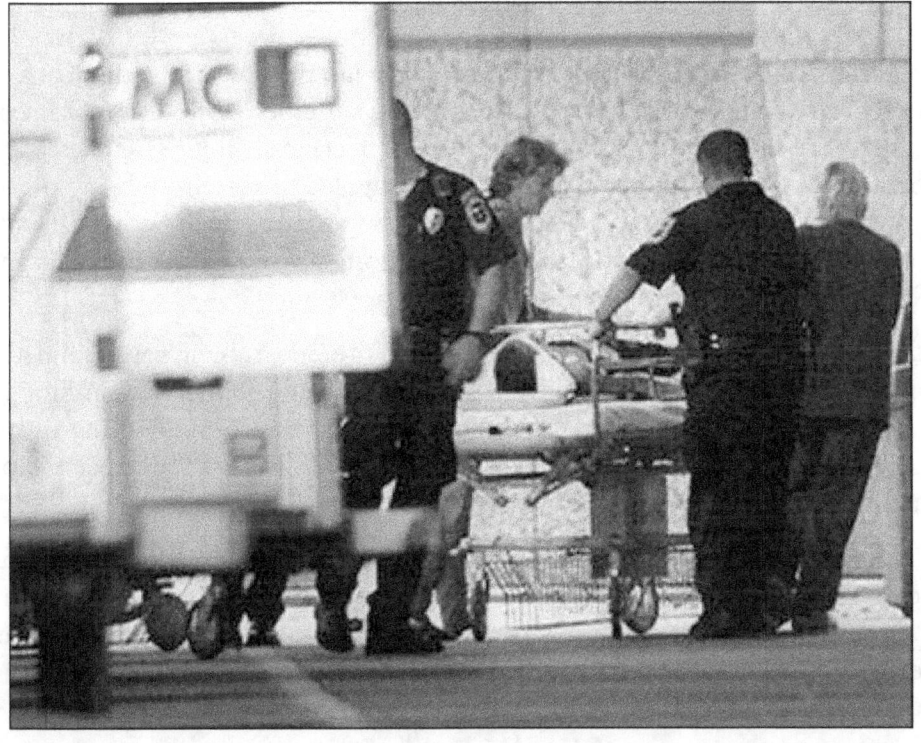

In the first 2 hours of the response, about 50 percent of the injuries were transported to hospitals by nonambulance vehicles—private vehicles and pickups. The scene required significant resources; sound triage decisions; and efficient oversight of EMS providers, firefighters, and citizen volunteers. Except for some transportation delays in the Northeast Division, loading and moving of patients proceeded efficiently. When ambulances arrived they received patients quickly and left the scene immediately. EMS providers followed MnNIMS guidelines and adopted good triage procedures.

Southwest Division: Medic 486 was assigned originally to 500 2nd Street, SE, but after hearing the status report decided to relocate to West River Parkway and 13th Avenue, S, on the south side of the river. Since they had responded from farther south in the city and had not tried to cross the river, they became the first EMS providers to arrive on the south side. They followed the Regional Incident Response Plan, established a Southwest Division, and assumed Triage Supervisor and Transport Supervisor positions. Fire activity and charged hoselines required Medic 486 to determine an alternate access and egress route quickly. They were inundated with the walking wounded and uninjured—many of whom were school-age students from a bus. The critical patients were triaged and transported promptly by local and mutual-aid services.

The ambulatory patients were kept in front of the Red Cross Building. Unfortunately, someone directed this group to a new location inside the Red Cross Building without consulting or notifying Medic 486. One of these patients was originally triaged as "delayed" but was considered "critical" by the time EMS was redirected to the new location. Moving lower acuity patients to a protected area is acceptable, provided that triage personnel are involved and are available to reassess them. In this instance the Division Supervisor had no knowledge of the move.

Southeast Division: The Southeast Division did not have any victims requiring EMS transport. EMS personnel stood by to provide rehabilitation and care for providers working in the area.

Emergency Medical Services Incident Command–EMS responders faced many difficulties managing the incident. The incident presented collapse with rescue, fire, water emergencies, significant public utility damage, infrastructure challenges, and the medical needs of critically injured and less injured citizens who were spread over many locations. EMS managers struggled to keep from being overwhelmed by these factors and the complexity of incident operations. Supervisors from both local and mutual-aid services responded, and were assigned positions as Liaison, Division Supervisors, Staging Manager, and Safety.

The principal EMS agencies in the metro area are not fire-department based, and typically they manage their daily operations with a single Operations Manager or Supervisor–calling in additional supervisors as needed. They are unaccustomed to operating in an expanded Command structure, and struggled to define the roles and duties of additional management staff once overall divisional Command had been established.

While EMS Incident Command worked to gain control of the overall situation, divisional operations were focused on rapid assessment, triage, and transport. Fortunately, with one exception, all patients transported from the bridge survived. This patient was triaged as critical, presenting in a trauma-induced cardiac arrest. Although the first to be transported, the patient had little chance for survival and died at the hospital.

Figure 10 represents the EMS ICS chart prior to 7 p.m. on August 1st.

Figure 10. EMS Incident Command System.

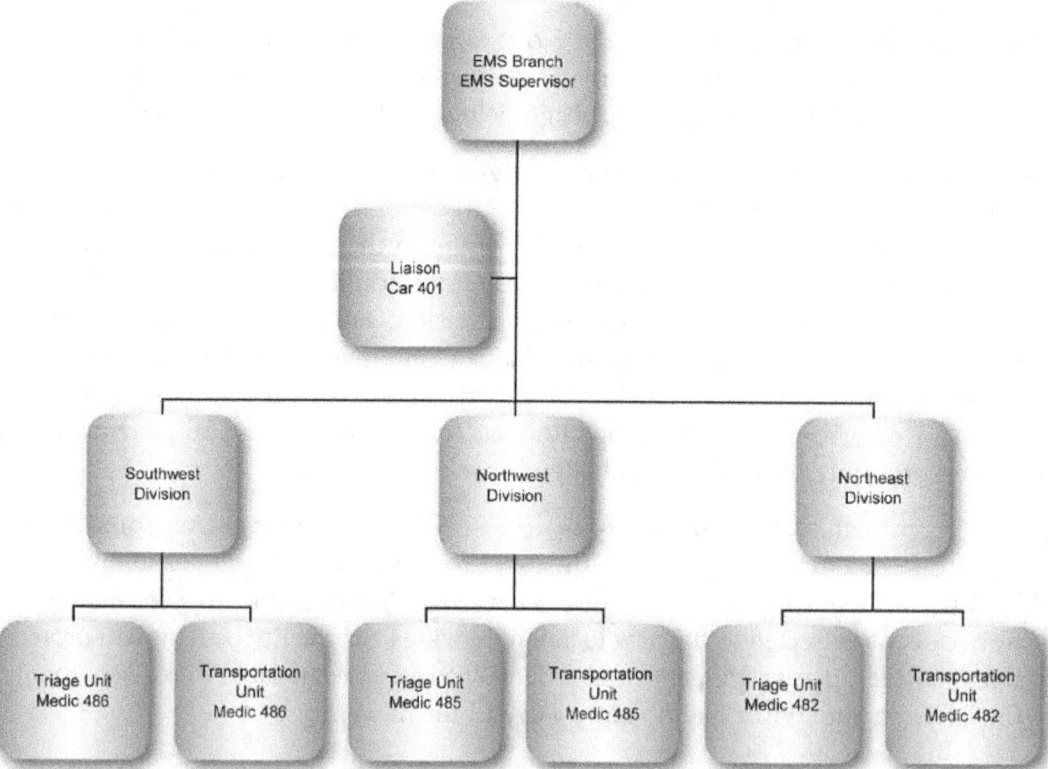

Emergency Medical Services Transportation–The incident occurred entirely in the HCMC EMS service area. Following their mutual-aid agreements, and using the Metro Region EMS Incident Response Plan (IRP), mutual-aid units from North Memorial Ambulance and Allina Medical Transportation provided significant response to the incident. Two nonmetro or out-of-State mutual-aid units, Kanabec County Ambulance from Mora, MN, and Lakes Region EMS from North Branch, MN, played key roles transporting patients. Both had completed patient transfers to hospitals in Minneapolis and were in the immediate vicinity of the bridge. Lakes Region Unit was assigned to the bridge by North Memorial Dispatch as part of their mutual-aid response. Kanabec County Ambulance was stuck in traffic trying to cross the I-35W Bridge where they were directed by law enforcement to assist EMS operations in the Southwest Division.

Between 6:05 p.m. and 8:11 p.m., 50 patients were transported by EMS to various hospitals. EMS units responding to the scene, Staging, or to cover the city are listed below (Table 2). This information illustrates how multicasualty incidents require management, provider, and logistical support to assist ambulance personnel.

Table 3: EMS Units Responding to the Incident.

Unit Type	Number of Units
Ambulances	31
Multicasualty Incident (MCI) Supply Trailers	1
MCI Supply Trailers (Support)	6
Supply Van	1
Command/Support Vehicles	7
EMS Incident Management Team Members	16

Within 90 minutes the regional EMS providers were able to staff 31 ambulances. Three hospitals implemented their disaster plans and three EMS EOCs were activated. The ability to mobilize EMS personnel throughout the community gives citizens confidence that EMS needs can be meet during large-scale incidents.

There were three patient collection areas at the incident and two transportation groups set up—one for each side of the river. During an MCI, West Metro Medical Resource Control Center (WMMRCC) is charged by the Regional Incident Response Plan with the responsibility to coordinate hospital destination and to track patients. A principle of managing MCI incidents includes the organized transfer of patients to the appropriate facilities within the appropriate timeframe. There was a breakdown in this function at the bridge because most units transporting patients did not communicate with WMMRCC even though dispatch, and EMS Branch gave multiple reminders on the radio.

EMS transported 50 patients to area hospitals. As mentioned, all but one survived. The non-survivor experienced cardiac arrest at the scene. Table 3 indicates which facilities received patients.

Table 4: EMS Transport of Patients on August 1st.

Hospitals	August 1st
Abbott Northwestern	5
Hennepin County Medical Center (HCMC)	24
North Hospital	9
University of Minnesota	9
Unknown	3
Total	50

Source: Kummer, C. (2007). *EMS and Hospital Patient Statistics.* (August 2, 2008). Unpublished manuscript.

Along with receiving patients by ambulance, hospitals also received patients by nonambulance vehicles during and after the first hours of the initial incident. On August 1st, 25 additional patients walked into various emergency departments, reporting they were injured at the I-35W Bridge (Table 5).

Table 5: Walk-Ins to Area Hospitals on August 1st.

Hospital	Number of Patients
Abbott Northwestern	2
Methodist	2
Ridges (Burnsville)	2
Saint Joseph's (Saint Paul)	1
Saint John's (Maplewood)	4
United	1
Unity	5
University of Minnesota	6
Hospital Unknown	2
Total Walk-In Patients	25

Source: Kummer, op.cit.

On August 2nd 35 additional patients sought emergency care at area hospitals or clinics, increasing the number of injured to 110.

Table 6: Patients Transported to Minneapolis Area Facilities on August 2.

Hospital	Number of Patients
Abbott Northwestern	0
Hennepin County	3
Methodist	2
Minneapolis Children's	11
North Memorial	6
Saint John's	1
University of Minnesota	12
United	0
Total	35

Source: Burfiend, M. (2007). *Hospital Compact Metro Region.* (August 2, 2007). Unpublished manuscript.

The I-35W Bridge incident ultimately claimed 13 lives and injured 110. By August 7th, there were still 10 patients in the hospitals.

Emergency Medical Services Communications Center–EMS communications was the responsibility of the Hennepin County EMS system, which operates separately from fire or law enforcement dispatch. The regularly assigned Emergency Medical Dispatcher (EMD) staff (three persons) was quickly augmented by an additional EMD present in the Dispatch Center. The additional EMD was assigned to the incident, with the others continuing to receive and dispatch the calls not related to the bridge incident. The Communications Supervisor was still on campus and immediately returned to the center. A fifth EMD who had just left work returned to the center and assisted with EMS communications.

The Communications Supervisor assigned an EMD to each of the following duties:

- normal EMD radio activities;
- backup EMD, Medical Priority Dispatch (EMD)/Pre-Arrival Instructions (MPD/PAI);
- I-35W Bridge incident;
- WMMRCC normal operations; and
- WMMRCC for the I-35W Bridge incident.

Along with incident communications, there was increased activity outside of the communications center. The communications staff often were asked for information, and noncommunications personnel entered the hallways. Superfluous noise made it difficult for the EMDs to monitor multiple radio talk groups. During the incident, EMS communications missed a message from Minneapolis Dispatch warning of possible hazardous materials on the bridge.

There was a delay in getting additional EMS supervisors to the scene due to their location at the time of the incident. For a while communications center personnel had to assume a major role in directing the EMS response[3], according to a Minneapolis document which noted

> ...the Hennepin EMS Communications Center assumed a prominent role in reaction to the escalating response. This proved to be both beneficial and detrimental to the EMS Branch Director. Beneficial in that the Communications Center was able to monitor and track information more effectively than the EMS Branch Director; detrimental in that during the evolution of the event, the EMS Branch Director was often in competition with the Communications Center for radio time...

Communications personnel noted that it would have been impossible to manage the center without the additional EMDs. The supervisor considered suspending MPD/PAI procedures, but determined that making another change from normal operations would worsen confusion. The supervisor's decision likely was correct, since properly used MPD/PIA may be even more important during system surge.

West Metro Medical Resource Control Center is fully integrated into the Hennepin County EMS Communications Center and is managed by the EMDs. Serving as the *medical command* for Hennepin EMS, it monitors MNTrac, the hospital closure system, managing and tracking patient transport destinations, and helping to control surge capacity. Even though this system exists, most everyday medical communications are conducted by cell phone. EMS personnel advised that most emergency department physicians prefer cell phone contact since they can answer requests for consultation from anywhere in the emergency department. When the resource center was needed, the EMS providers at the bridge continued to use cell phones instead of the WMMRCC. The result was hospital confusion, inability to track patients, and a delay in determining where everyone was transported. Hospital emergency departments also failed to use MNTrac consistently, leading to confusion as to surge capabilities at emergency departments.

---

[3] Nonmunicipal EMS systems tend to field fewer EMS supervisors for daily operations.

CHAPTER III. LAW ENFORCEMENT AND THE FAMILY ASSISTANCE CENTER

The Minneapolis Police Department (MPD) and the Hennepin County Sheriff's Office were the primary law enforcement agencies responding to the disaster. MPD eventually would assume Incident Command from the fire department once the initial rescue effort was completed. Police had jurisdiction on the land side, while the Sheriff's Office had authority over water-based operations, per State statute.

The MPD set up the law enforcement Command Post (CP) at 12th Street River Parkway in a parking lot that served the city's Red Cross headquarters and a commercial establishment (see Figure 11). The first critical tasks were establishing and controlling a very large and physically challenging perimeter, controlling intelligence information, and initiating an investigation to rule out the possibility of terrorism.

Figure 11. Minneapolis Police Command Post.

The prospect of terrorist involvement was a very real concern. In 2006, the Minneapolis Federal Bureau of Investigation (FBI) Office completed a Domestic Terrorism Threat Assessment and identified special interest, left- and right-wing groups, and lone-wolf actors. The Minnesota Patriots, a militia group, were involved in a high-profile incident in the 1990s, manufacturing Ricin for use against local law enforcement. Timothy McVey was known to have checked out the Federal building here, before eventually deciding on the Murrah Federal Building in Oklahoma City. In addition to domestic terrorist groups, Zacharius Moussaoui and other high-profile international terrorists were arrested within the Twin Cities metropolitan area. The local FBI Joint Terrorism Task Force (JTTF) has been among the most active in the Nation, addressing the issue of overseas financial transfers and groups such as al Qaeda, Hizballah, Hamas, al-Ittihad al-Islami, and Islamic Jihad. The Twin Cities area also is home to many high-tech defense contractors, corporate headquarters, and other critical infrastructure.

A police inspector received a text message about the bridge collapse and was detailed as the investigation chief. He also was put in charge of opening a Family Assistance Center (FAC). The inspector went immediately to the 10th Avenue Incident Command Post (ICP) (as he had learned from the National Incident Management System (NIMS) training) to be briefed on the status of rescue efforts. He quickly paged out the special weapons and tactical (SWAT) team for rappelling gear. The Federal Bureau of Alcohol, Tobacco, Firearms and Explosives (ATF) arrived shortly thereafter. The Saint Paul Police Department assisted MPD by handling the mutual-aid calls. The U.S. Secret Service did tactical mingling and asked how they could help.

The Investigation Chief for the incident also directed a Special Operations Commander to ascertain whether an explosion caused the bridge to collapse. It was crucial to determine whether the disaster was a criminal or terrorist act or an accident. The Special Operations Commander called out metro-area Explosive Ordnance Device (EOD) Teams per established mutual-aid agreements. He requested a mobile Command vehicle and homeland security staff as well. The Commander had headed up a detail that responded to Hurricane Katrina and had learned first-hand what was necessary to maintain an operation over an extended period of time. For example, he had worked already with the local Target store to identify and prestage crisis supply requirements, e.g., bottles of water, sunscreen, cots, generators, message boards, and other supplies. A phone call to Target activated these standby supplies immediately. The store sent an asset control team to help with requirements size up, and 45 minutes later they delivered the first of several pallets of supplies. The planning done ahead of time made a big difference in how quickly these supplies were mustered.

The MPD Chaplain arrived and was asked to work on arrangements for the FAC. He coordinated site location and staffing arrangements with the Red Cross, the City's Department of Health and Family Support, and relevant Hennepin County offices. Teams of officers were sent to hospitals to follow up with the injured, who had been transported to eight different medical facilities.

The three key law enforcement commanders were skilled in strategic planning. All had been trained in SWAT and in homeland security. They had trained with other Minneapolis first responders and with metro area and State law enforcement agencies, as well as with Federal organizations. Most of the major stakeholders knew each other, which made communications smoother and facilitated cooperation.

MPD set up 12-hour platoons and used a helicopter to obtain a larger view of the scene and traffic. They had to determine where to put up cement barriers and where to use police officers instead to control access. The perimeter they established was big—2-1/2 square miles, and was very labor-intensive to maintain. The traffic and perimeter control division also handled all dignitary visits and assigned officers to escort Minnesota Department of Transportation (MNDOT) officials and construction company workers to ensure the integrity of the scene. They strictly enforced a rule prohibiting taking photos or "souvenirs" so that the scene was not disturbed.

The Saint Paul and Bloomington Police Departments handled CP security, and the Minnesota State Patrol managed all the other mutual-aid staffing and coordination.

Minneapolis opened a FAC at the Holiday Inn-Metrodome near the incident site within 2 hours of the bridge collapse (see Figure 12). The MPD, the American Red Cross, and the Minneapolis Department of Health and Family Support (MDHFS) shared responsibilities for operating the FAC.

Figure 12. Site of First Family Assistance Center at the Holiday Inn.

The FAC remained at the Holiday Inn from Wednesday, August 1st, to Friday, August 3rd, when it was moved to Si Melby Hall at Augsburg College. The move was necessary because the hotel needed the space for a previously scheduled event.

The Health Commissioner activated mutual aid with public health departments in the metropolitan area to secure additional mental health providers. Volunteers from the Medical Reserve Corp also were activated, as were police chaplains regionally through mutual aid activated by the MPD. Center personnel provided psychological first aid and current information to the families of the deceased and the missing. Staff and volunteers at the FAC were committed to providing an environment where family members could feel safe, communicate information, receive comfort, and connect with friends and loved ones. Families also received guidance on services that would be helpful. It was at the FAC that missing person information was collected and inquiries were made. The FAC was the main location for formal briefings and status reports, which were delivered by MPD, the Hennepin County Sheriff's Office, the Hennepin County Medical Examiner, the FBI, and the National Transportation Safety Board (NTSB).

The FAC area at the Holiday Inn was declared a media-free zone, and many measures were put into place to ensure privacy for the families. This was particularly important because many members of the media were staying at the same hotel. At the Augsburg College site, private rooms were set aside for the MPD chaplains to handle death notifications.

There were benefits to having a mix of agencies and their respective services immediately available at the FAC when it opened, and the services were well executed. However, there also was uncertainty over which agency had the lead for direction and decisionmaking, because the details for partnership and cooperation had not been scoped out in advance. The Minneapolis Emergency Operations Plan does not cover a FAC. Indeed, Federal guidelines, including NIMS and the National Response Framework (NRF), do not address the requirements, coordination essentials, and lead agency options adequately for FACs.

Much credit goes to the MDHFS for completing its own after-action review of the FAC after it closed on August 11th. In that review, MDHFS details a thorough and honest assessment of both positive features and problems encountered. All agencies involved with the FAC effort were given an opportunity to provide input and ideas for improvements. Changes already have been made based on their critique, which serves as an outstanding example of how to document the lessons learned and use them to make important policy and procedural changes.

Some of the lessons that Minneapolis learned, and their recommendations for a FAC, are presented below because they can enlighten disaster planners everywhere as to some of the important issues.

1. In choosing a FAC location, make sure that it offers, among other essentials, private space[4] for sensitive communications with families.

2. Food services should include options that are sensitive to the population served, once that census can be determined (e.g., vegetables, no pork, etc.).

3. Staffing protocols should be developed to provide consistency and ensure smooth transitions.

4. Scribes are an important asset and should be scheduled so that this position is covered until the FAC closes.

5. Activation procedures for a Medical Reserve Corps and other health and mental health resources should be clear.

6. Increased training on the elements of psychological first aid (PFA) would be beneficial, as would clarification of how PFA, crisis counseling, and critical incident stress management are different, and how to know which is most appropriate under the circumstances.

7. Some basic recordkeeping forms were used, but additional ones would have been helpful. The after-action report recommends that more basic recordkeeping templates be developed and then used immediately upon opening a FAC.

8. Staff surge is needed early to set up the site, establish recordkeeping and office procedures, and cover clerical tasks.

9. Whiteboards and chart paper with easels are essential supplies.

---

[4] Note from USFA, if the FAC is in a hotel, avoid planning to use guest rooms as private counseling space; they are not appropriate.

10. Identify beforehand what points should be covered in briefings at the change of shifts.

11. Plan for an adequate number of land-based phones and computers in the first days of FAC operation. Check on the actual accessibility of printing and Web services at designated FAC locations.

12. FAC staff members, including mental health professionals, typically are not as familiar with the ICS structure as are first-responder agencies. This lack of awareness can cause delays in communications and in following orders. This can be addressed through training and clear chain-of-command protocols.

13. In an effort to provide emotional support to families, it is possible to mobilize too many mutual-aid organizations for mental health support versus other health or information skills that also are needed. Jurisdictions should develop plans to determine the types of volunteer services needed and estimate how many individuals would be required. Use only as many sources as are needed.

14. Remember that FAC workers need emotional support and encouragement, too. Make plans to provide care for the caregivers.

CHAPTER IV. RECOVERY OPERATIONS AND THE MEDICAL EXAMINER'S OFFICE

The Hennepin County Sheriff's Office (HCSO) has responsibility for all water rescue, body recovery, and removal of hazards to navigation in the city and county. The office has used Incident Command at literally hundreds of water incidents that have occurred on the 103 lakes and three rivers located within the county. Since 1967, Hennepin County has maintained a mutual-aid agreement with all law enforcement agencies in the area. The agencies have an annual refresher course to update the resource list and review procedures. The Sheriff's Office also has a good working relationship with the U.S. Army Corps of Engineers, the U.S. Coast Guard, and local fire, emergency medical services (EMS), and emergency management agencies with which they have responded on incidents over the years.

The HCSO Supervisor was on the water side at 6:26 p.m. and became the Water Rescue Group Supervisor. He staged water rescue and recovery operations at the University of Minnesota's river boat flats, about a quarter mile from the disaster site. The Supervisor designated a Staging Officer and a Safety Officer. All personnel were well equipped. If they did not have a radio, they were given one. The HCSO Water Patrol has a logistics trailer and a generator that they can hook up to their Command Posts (CPs) and into other CPs. Forfeiture funds and grants had been used to purchase side-scan sonar, which the divers used to read the images in the water and then mark the locations with Geographic Information System (GIS).

Figure 13. Hennepin County Sheriff's Office Personnel at Water Command Site.

The Supervisor sent a notification dispatch for Hennepin and Minneapolis Police Departments (MPD) using an intrastate radio system. Five counties also sent help. The Supervisor was able to access the pool channels right away; they were clear, and users experienced no problems. There were frequent communications between the Water Rescue Supervisor and the Minneapolis Fire Department's (MDF's) Incident Commander (IC).

Twelve agencies with 28 watercraft arrived within the first hour. The U.S. Army Corps of Engineers' St. Paul District's Maintenance and Repair Unit supported recovery operations by providing a floating plant, a crane, and boat operators. They also placed liaisons in the State Emergency Operations Center (EOC) and at the Unified Command Post. The National Guard made available a helicopter equipped with a hoist.

The U.S. Coast Guard (USCG) Upper Mississippi River Sector, which includes the ports at St. Louis, Kansas City, and the Twin Cities, had been working with their Area Maritime Security Committee on plans and table exercises related to preventing and responding to potential terrorist events. This training familiarized them with their particular role as a support resource to other levels of government. After the bridge collapsed, USCG sent a reservist to serve as the USCG liaison at the State's EOC. Two 25-foot boats with crews were diverted from near Omaha and sent to Minneapolis, arriving within 5 hours. Other boats were sent from their location at Saint Paul. They assisted with the search and rescue in conjunction with the HCSO Commander. Noted the USCG Captain, "We respect that locals are the lead on response."

By 7:27 p.m., it was determined that all individuals on the bridge and next to the water had been rescued. The recovery phase officially began at dawn on August 2nd.

In preparation for recovery operations, fire and rescue personnel were demobilized and debriefed. The Sheriff's Office tightened the perimeter and established a 6-mile patrol on the surface and the shore, looking for survivors and evidence. The U.S. Coast Guard closed traffic on the Mississippi from mile marker 847 to mile marker 854, establishing this section as a safety zone. The zone essentially encompassed the section from the Ford Lock and Dam (downstream) to St. Anthony Falls (upstream). The Hennepin County Dive Master was placed in command of dive operations, and all other divers worked under his orders. Early in the situation two dive teams that did not follow orders had to be removed from the water. A rule was set in place that only public safety divers would be permitted to engage in dive activities. Those divers were used to diving in currents using the same equipment and had trained and worked together with underwater communications. They were accustomed to swift-water rescue and salvage work.

Dive operations were set up in 12-hour shifts during daylight hours. Each morning at 8, divers were debriefed from the previous day's progress and new goals and objectives were set covering bank to bank and 100 yards inland. Before dives can be executed, many setup tasks must be handled, mostly to ensure the safety of the divers. The hazards divers faced included

- no visibility (pollution, the impact of heavy rains, etc.);

- a 3-mile-an-hour current that increased when storms developed later;

- early threat of additional collapse of the bridge structure;

- sharp metal and concrete debris;

- bloodborne pathogens;

- gas and oil from the vehicles in the waters;

- leaking railroad cars on the north bank; and

- threat of lightning and thunderstorms.

The U.S. Army Corps of Engineers brought in a barge and a 1-ton crane to assist with water operations. The barge was used as the diving platform (Figure 14). The local dive teams were joined by the Federal Bureau of Investigation (FBI) Underwater Search and Evidence Recovery Team until Sunday, August 5th. About 20 divers from Hennepin and Dakota County Sheriff's Offices plus the Hudson Dive Team (police, fire, and EMS that specialize in river diving) worked upstream from the collapse area. As much debris as possible from the broken bridge had to be cleared out in order to get to victims who remained inside submerged vehicles.

Figure 14. Divers Operating from the Barge.

For every diver who was in the water, the HCSO had two on the land as backups and water rescue teams, much in the same fashion as fire department Rapid Intervention Teams (RITs) stand by during structure fires. Land personnel released 1 foot of line at a time to the divers so they could move slowly through the murky, debris-laden water. Divers used the sonar that had been purchased through homeland security grant funds to locate debris, vehicles, and human remains—the locations of which they marked using GIS. Divers identified license plate numbers and found other evidence that helped support identification of the deceased.

President George Bush came to Minneapolis on August 4th and toured the site of the bridge collapse. During his visit he asked local officials if there were any resources they needed. A request was made for Navy divers, whose equipment and more intensive training were needed to accomplish recovery more quickly. Within 12 hours from the time of the request, a Navy Commander was on the ground, followed soon thereafter by the U.S. Navy Mobile Diving and Salvage Team #2.

Figure 15. Navy Divers.

On August 5th, the Water Group Command site was moved to the Ford Lock and Dam. The divers, the MFD, and the U.S. Army Corps of Engineers all worked under the same Water Rescue Group. Heavy rescue assets from the fire department helped the Navy dive team, which began diving on Monday, August 6th. The U.S. Army Corps of Engineers got a contractor with commercial divers to help, and the National Oceanic and Atmospheric Administration (NOAA) emailed hourly weather reports so ICs could stay on top of potentially hazardous weather. (Dive teams and cranes need 30 minutes to get out of the water and shut down.)

On Saturday, August 11th, a major weather front arrived, bringing heavy thunderstorms that wiped out dive operations and affected security posts, the CP, and the barge. The dive sites had to be rebuilt. The current rose from 3 miles an hour to 7 miles an hour, making it impossible to continue diving. At the same time, local officials were under pressure from the Port of Minneapolis to re-open the river so that barge traffic could recommence in support of upstream facilities and businesses. The heavy rains had brought dive operations to a virtual standstill, and created poor conditions that would take days to resolve, thus further delaying re-opening the affected part of the river. In response, the U.S. Army Corps of Engineers worked with local leaders on a plan to control the water level. From the Coon Rapids Dam they lowered the water by 2 feet for a 12-mile stretch. That, in turn, made it possible to lower the water level by 5-feet for every one-eighth mile between Upper St. Anthony Falls and Lower St. Anthony Falls, the area in and around the dive site.

When Navy divers found human remains that were outside submerged vehicles they bagged the remains underwater and then brought the bag to the surface on the left descending bank of the river (near the University of Minnesota research buildings). Vehicles also were tarped underwater before being brought to the surface, as most still contained human remains. Two of the bodies had to be recovered in portions over several days. It took 20 days until the eighth and final victim was recovered. All told, 58 divers worked over 5 days to accomplish debris removal, evidence collection, and body recovery.

The Coast Guard brought a 25-foot response boat upriver from Saint Paul, plus four additional boats. Crews from the bridge construction company that was working on the bridge deck prior to the collapse helped to remove construction debris. The Minnesota Department of Transportation and personnel from the Minneapolis Department of Public Works and engineering office also were involved and participated in briefings. A site called the Bohemian Flats became the designated location for dumping pieces of the bridge and deck that were salvaged from the river (Figure 16).

Figure 16. The Bohemian Flats Salvage Site.

Local officials maintained a Sheriff's Captain at the Minneapolis EOC and the Hennepin County EOC until water operations were final. The water river rescue group functioned as a major part of the Operations Section of the incident.

MEDICAL EXAMINER'S OFFICE

The Minneapolis 9-1-1 Communications Center contacted the Hennepin County Medical Examiner's (HCME) Office to report the incident and alert the office to fatalities from the scene. The HCME is staffed with four full-time physicians and operates around the clock. The Chief Medical Examiner went to the EOC the first night, and the HCME office was filled, almost to excess, as many staff had been called in to handle the flood of phone calls from media around the world.

There were several priorities that were taken care of at the outset. First, HCME asked the MPD's Homicide Division, specifically the Police Chaplain, to handle all the death notifications to the families. The Medical Examiner's Office maintained strict control over victim information, ensuring that families were the first to know the status of their loved ones. HCME also contacted the Ramsey County Medical Examiner's Office with which they have a mutual-aid agreement, and asked them to assume that if they had water victims, those bodies in all likelihood would be from the bridge collapse. Ramsey County is downstream from the site of the accident and HCME wanted to make sure that all victims from the disaster were handled through the same medical examiner's office for continuity with the families and with the investigation process underway. Dakota County also went on standby, and the Minneapolis-Saint Paul Airport took action to ready their morgue according to plan. There were fewer fatalities than initially anticipated, so HCME had the capacity to handle all the casualties.

FATALITIES

Three fatalities were recovered early from the scene; another individual was administered cardiovascular resuscitation, but died; and a fifth victim in grave condition was transported to the hospital, but succumbed to the injuries. An additional eight bodies were recovered by the special dive teams from the Mississippi River over a period of 20 days.

Victims from the bridge disaster were identified through a combination of fingerprints, drivers' licenses, dental records, and in one case, DNA. The MPD's laboratory produced a report on the DNA sample within 3 days.

HCME's procedure was to have the police chaplains notify the families once a presumptive identification was made. Families were told that it would take a few hours or longer to make a scientific identification, after which the remains could be released to funeral homes. Once scientific identification was confirmed, the HCME's office and the chaplain both met with the family. All the deceased victims were from the local area.

ANTEMORTEM AND POSTMORTEM DATA

Until midafternoon on Friday, August 3rd, the Family Assistance Center (FAC) was loosely organized, with several offices sharing Command. There were some problems with recordkeeping at the FAC for calls from people who were concerned about missing relatives and friends. Detailed information was needed to help identify the victims, but that information had not been thoroughly documented. The HCME is authorized to obtain some information but consulted with the County Attorney to ascertain whether they could initiate calls for information and make better progress on presumptive identification.

Teams of law enforcement officers were sent to hospitals to identify and follow up with the injured who had been transported. They needed to document who was injured, who was missing, and who was dead. Within 14 hours the identities of all but eight individuals who were missing were confirmed. Between August 1 and August 10, officers tracked down more than 1,200 leads that were called into the MPD, the National Transportation Safety Board (NTSB), and the Red Cross on missing persons. Homeless individuals were known to camp under the bridge, so there was concern that possibly one or several homeless were among the missing. The police worked with a homeless man who knew that area and they were able to rule out any homeless people who frequented that area as being among the casualties.

The HCME noted that antemortem and postmortem data were not linked quickly enough. This is a common challenge during mass casualty incidents. Contributing to the problem is the fact that different agencies want different sets of information, and families often have to go through multiple requests for information. For example, State Police collect information that is different from what the NTSB requires, while medical examiners want only information that helps to answer questions about cause and manner of death.

The Medical Examiner would like to see the development of a unified data collection form, whereby essential information about a missing person, or someone known to have perished, can be entered into a computer and saved. The data could be updated or corrected later, as needed. A law enforcement officer or a FAC staff person could sit down with the family and collect the information directly. That data then could be compared against any antemortem data collected and stored by the medical examiner or coroner and analyzed for possible matches. HCME is hoping to use a State grant to create a prototype of this system.

Another recommendation from the after-action report was that medical examiners and coroners should consider setting up a Web site or be linked to the main government Web site if a mass-casualty disaster occurs. Periodic press releases and other information that can be made public could be posted there, and callers could be directed to go to the Web site for the most current information. This idea is one solution to the problem of being inundated by calls for information, and having to satisfy those requests without sacrificing time needed to work directly with the families and the victims. Once recovery operations began, Minneapolis responders directed all media inquiries to the Medical Examiner's Office.

RESPECT AND SENSITIVITY TO VICTIMS' FAMILIES

The dive operations were closely coordinated with both the HCME and the MPD. A decision was made to shroud all the vehicles brought up from the water. The Medical Examiner's personnel conducted a brief investigation inside each vehicle at the river bank. The shrouded vehicle then was loaded onto a flatbed truck and escorted by police to the coroner's garage for further processing. All vehicles from that garage were removed in anticipation of the arriving bridge collapse victims. There was security at that site, and the area was covered over to prevent any public viewing. Whenever a victim was recovered and identified, the family was called to the FAC for formal notification and support. At all times, the privacy of those who had lost a loved one was protected and maintained.

As another example of sensitivity to the families, on August 9, the Augsburg Drum and Bugle Corps was scheduled to be filmed performing in the athletic field located to the south of the FAC (the center had been transferred to Augsburg College several days after the bridge collapse). The families were made aware that filming would be taking place and were assured that they would not be filmed or identified.

CHAPTER V. MINNEAPOLIS EMERGENCY MANAGEMENT SYSTEM

Emergency management in Minneapolis is handled by the Office of Emergency Preparedness (OEP), an agency separate from traditional fire and police agencies. By city ordinance, OEP has responsibility and authority for overseeing the city's response under a Declaration of Emergency. The OEP is also responsible for preparedness, mitigation, recovery, training, and grant administration involving emergency preparedness. OEP operates under an all-hazards approach, and conforms with the Department of Homeland Security (DHS) National Incident Management System (NIMS) standards.[5]

The OEP is staffed by a director who reports directly to the city coordinator and Mayor. He is assisted by two deputy directors, and various staff members, many with emergency management certification and public safety experience. Community partnership is an important guiding principle for the OEP. Minneapolis residents, businesses, and property owners are considered major stakeholders of OEP. This was demonstrated during the bridge incident as several private companies including the Target Corporation, Minneapolis Queen Riverboat, river services, rescue product providers, food suppliers, and dry goods suppliers called 9-1-1 to volunteer services and equipment. After the Emergency Operations Center (EOC) opened, these calls were forwarded to the Logistics Section.

The OEP also considered partnerships with community, State, and Federal partners an important element of emergency management. Training and exercises involving State, county, and city agencies also helped build the working relationships evident during the incident. Key city and Hennepin County elected officials, and staff attended the Emergency Management Institute (EMI) Integrated Emergency Management program in Mt. Weather, Virginia, and later conducted a gap analysis of the city's emergency response plan. Many local officials pointed to this experience as key to developing Minneapolis' current emergency response plan. Further details are described later in this report.

MINNEAPOLIS EMERGENCY OPERATIONS CENTER

The City of Minneapolis, Hennepin County, City of Saint Paul, Ramsey County, and Hennepin County emergency medical services (EMS) all opened EOCs shortly after the bridge collapse. The State of Minnesota and the Federal Emergency Management Agency (FEMA) Region V (Chicago) also opened EOCs within 2 hours. The Minneapolis EOC was the lead, and acted as a Multiagency Coordinating (MAC) Center. This report will focus on the Minneapolis EOC, with additional information on how the other EOCs provided assistance. Public and governmental agency support indicated that the breadth and depth of response was good, with only a few minor system glitches.[6] Contributing to the mission's success was the fact that representatives from all 10 NIMS disciplines were in the EOC.

---

[5] City of Minneapolis. (n.d.). *Office of Emergency Preparedness: Organizational Command Structure.* Unpublished presentation.

[6] Lee, C., and Lewis, P. (August 3, 2007). *With Minor Exceptions, System Worked.* Washington Post.com. www.washingtonpost. com/wp-dyn/content/article/2007/08/02/AR2007080202262

The EOC is located in the basement of city hall and is used when a large-scale emergency or disturbance occurs that involves multiple city agencies. The EOC is essentially a single room, which did not have enough space for all of the representatives from the organizations having statutory authority to be present. There is not enough room for a policy coordinating group, usually staffed by political and administrative leaders, or for other planning, logistical, and public information functions. The inadequate size and functionality of the current EOC was cited by most respondents as the biggest obstacle in the management of the response. Particularly during the evening of the collapse, the EOC was simply not capable of handling the number of staff and elected officials who reported to the center (Figure 17).

During the I-35W Bridge incident, the EOC was opened at 6:20 p.m. and staffed 24 hours a day for the first 4 days. Beginning August 5th, the EOC operated for 12 hours a day until closing on August 20th, when the last body was recovered.

Figure 17. Minneapolis EOC.

ROLES AND RELATIONSHIPS

A key to successfully managing major disasters is the predesignation and understanding of roles, and the development of relationships prior to an incident. These important elements were strengths in the Minneapolis response, as most participants from Hennepin County and the State had worked on or participated in, exercises with their Minneapolis counterparts. Time after time, the individuals interviewed for this report cited relationships as the single most important reason why the response went as smoothly as it did.

The Minneapolis EOC functioned as the primary EOC and as the MAC, coordinating the roles of other EOCs throughout the community (Figure 18). EOCs in Hennepin County, Hennepin County Medical Center, Saint Paul, Ramsey County, and the State of Minnesota were operational, some of which had specific responsibilities. For example, Saint Paul assisted with coordination of mutual-aid, and Hennepin County assisted with logistical needs.

EOC leaders and Unified Command personnel worked together closely. In many respects the most seasoned Incident Commanders (ICs) and department directors were available and responded on the night of August 1st. These leaders had planned, trained, and responded together on smaller incidents before, so they knew each other and the capabilities of their respective organizations. Turf battles, not uncommon in events of this size, were not a factor because of the relationships that had been developed over the years (Figure 18).

Figure 18. EOC and Unified Command for the I-35W Incident.

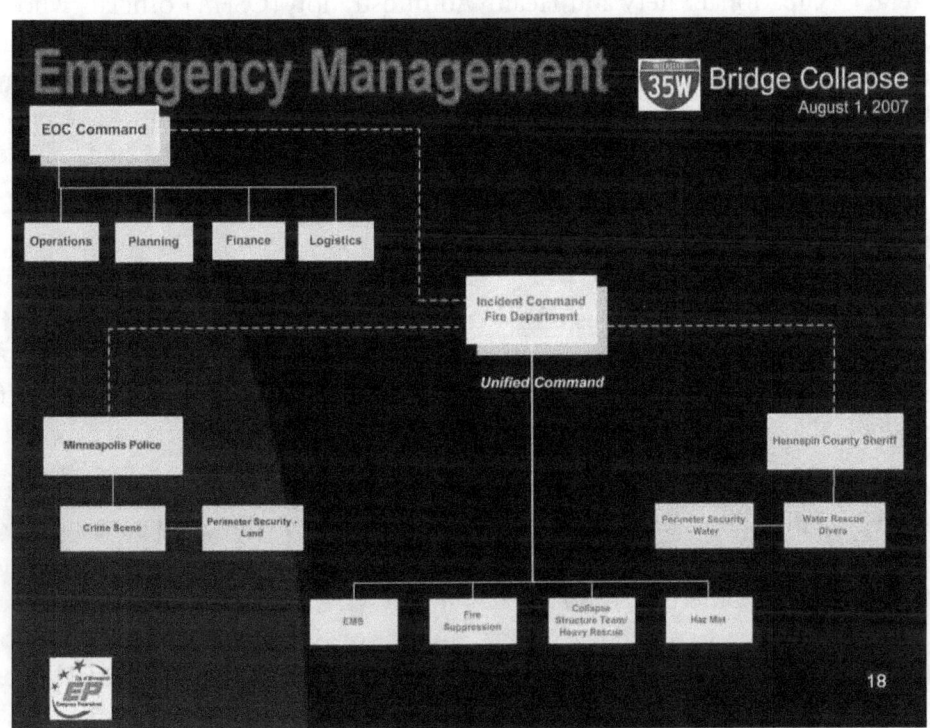

EMERGENCY OPERATIONS CENTER OVERSIGHT

Management of the EOC was the responsibility of the city emergency manager. His Command and General Staff included personnel with extensive emergency services experience in multiple venues. The EOC had primary teams, and plans were in place to ensure that position depth was maintained. Most city department heads and their ranking officers participated in EOC activities at different times during the incident. Fire, police, public works, communications, public information, and finance rotated trained personnel throughout the early part of the incident. Hennepin County and State Homeland Security and Emergency Management representatives always were present as well. When the EOC shifted to daytime operations, most agencies assigned one representative to cover the EOC.

Notwithstanding the limitations of space, the EOC was operational within 20 minutes of the collapse, and was well-managed. Personnel were able to focus on their specific duties despite interruptions and distractions. EOC participants adhered to NIMS and MnNIMS standards. They also had to manage Continuity of Government (COG) and Continuity of Operations (COOP) tasks. It appears that EOC personnel also handled those aspects well.

SAFETY

The overcrowding at the EOC could have been a safety issue. This problem was recognized by city officials and is being taken into account in planning for future facilities. There were safety activities on the scene, but no designated Safety Officer or Assistant Safety Officers. The EOC should have prompted Command to establish a formal Safety Group.

Fortunately, experienced fire department Safety Officers were on the scene the first night, and officers from several agencies helped to ensure a safe operation. Minneapolis officials were assisted by State and Federal Occupational Safety and Health Administration (OSHA) officials who contributed to safety monitoring. Rather remarkably, there were no reported OSHA accidents or injuries during the response and recovery periods—only a few minor injuries, none of which required hospital admission. This record is especially impressive considering that first responders were dealing with a combination of fire, fast water, tangled steel, suspended concrete, and crushed vehicles, and that Minneapolis estimated there were over 100,000 hours clocked on the scene.

PUBLIC INFORMATION OFFICER

The city Communications Director is usually designated as the Public Information Officer (PIO) for the EOC and, when indicated, is charged with establishing a Joint Information Center (JIC). Unfortunately, there was a delay in notifying the PIO, due to technical and human factors. The Deputy Emergency Manager was an experienced PIO and could step in to handle the situation.

After the PIO was located, the process continued, and efforts were made to coordinate information between the EOC and the PIO. As the incident developed, the PIO maintained consistent communication and coordination with other local and state public information officials, and established a central location for holding press conferences and media interviews. A critical event requiring coordination was the Mayor's 9 p.m. news conference where the Mayor, Governor, and senior public officials formally addressed the public. Well-coordinated major press conferences are critical to assuring the public that the situation is under control and public threats are mitigated. The press conference was a great success and set the stage for how the press and public would view the response from that point on.

The following day, the EOC Manager directed that press conferences be held at specific intervals. Status updates were released at 11:15 a.m., 1:30 p.m., 2:45 p.m., and 6 p.m. On August 3rd a similar press release schedule was followed. Also, by 5:30 p.m. that day, the city released the first official situation report, which continued through August 10th. The press briefings were important because they provided a framework for managing the requests from local, State, and international media. In a major disaster, most local governments are caught off guard by the sheer volume of press requests. Handling the media is a key element of effective emergency management, and that means setting up a robust public information capability.

As the rescue phase began to wind down, a JIC was formed and public information activities switched to the recovery process. These activities included the coordination of a Presidential visit, Governor's visit, a telecommunications briefing with legislators, Federal assistance from major government agencies, and community events.

LIAISON

A Liaison Officer was appointed to provide service between the EOC and the incident, and the EOC and political leadership. Good communications were critical at the EOC since the facility's physical limitations precluded many from being involved directly at that location. The Liaison Officer's job was made easier, since political and administrative leaders were trained in either IS-700 or IS-800 and were familiar with, and had confidence in, the emergency management team.

The EOC Liaison also assisted the PIO in managing the media, especially as national media and VIPs began to arrive in Minneapolis. The Liaison Officer assisted political officials in preparing for news conferences and other media events. Another aspect of liaison involves the 9-1-1 Center. Many calls were received offering assistance and materials. Some requests were forwarded to the EOC for a recommendation.

OPERATIONS

The EOC Manager appointed Section Chiefs for Operations, Planning, Logistics, and Finance/Administration. The EOC and the Section Chiefs worked together to implement the preplan for incident operations. There was excellent cooperation between the city and Hennepin County. The Operations and Planning Sections worked closely with Incident Command and responding organizations to determine when Command would shift to different organizations. There was smooth Transfer of Command from fire to police, and eventually from the city to the State.

PLANNING

The Planning Section Chief was responsible for short- and long-term planning for the incident. He prepared the Incident Action Plan (IAP) for the EOC, including its relationship to the incident and for the continuation of city operations. The Planning Chief's role was especially important in ensuring a smooth Transfer of Command from the city to the State.

The Planning Section Chief also was responsible for documentation of all resources used and needed. During the incident, some, but not all, of the NIMS documentation was used. Because of the limitations of the EOC, the initial planning information was recorded on an erasable white-board, then transcribed and photographed. Eventually the information was entered into a computer.

Planning for the assistance that other entities provide is a significant part of the planning process. There was close cooperation between Planning and Logistics, especially during the damage assessment phase of the incident.

Businesses were affected by the river closure and by the traffic detours around the site. Staff met with affected businesses to document how they were being affected and to explain the financial assistance that was available. Personnel helped assess financial damage and monitored economic dislocation and impacts over time. Sales and business traffic typically dropped by 20 to 40 percent for commercial occupancies closest to the collapse site. Reduced access to these locations was a primary problem. Staff also worked with Homeland Security and Emergency Management (HSEM) and the U.S. Small Business Administration (SBA) to process relief loans for small enterprises that were affected.

With the Chamber of Commerce, planners established a Web site that provided a communication tool for the business community to use in monitoring conditions in the days and weeks following the disaster. Staff assisted businesses that had to be relocated temporarily.

The city had to locate a site to which National Transportation Safety Board (NTSB) evidence and debris from the scene could be transported. Staff identified and secured temporary staging locations for this material.

Business Information Services and Technology–Various types of technologies contributed to response and recovery operations, from the side-sonar equipment to the 800 MHz radios and the wireless, bridge-mounted cameras. The city's office of Business Information Services (BIS), likewise, applied its technology tools to support the EOC members and ICs. Technology allowed the EOC to monitor the incident scene visually through video cameras that were placed at the site. The cameras provided real time, situational awareness of the site and assisted Command in understanding the incident's needs, and in communicating safety-related issues.

Using computer-based Geographic Information System (GIS) mapping applications, commanders and EOC personnel were provided a Web-based Common Operating Picture that included incident area maps (which helped responders establish and relocate perimeter security), access checkpoints, and Staging Areas with real-time changes that were made available to all city staff (Figure 19). Another application used was a Web-based traffic software management system that aided the public in identifying alternative routes and options for driving in and out of the city. This was a critical function because the I-35W Bridge was a major artery for access to and egress from downtown and was one of the busiest bridges on the Mississippi River.

Figure 19. Minneapolis EOC with Common Operating Picture.

The BIS Department also collocated a GIS work unit and a printer at the Minneapolis Police Department (MPD) Command Post (CP). Maps were supplied to mutual-aid responders, including the Secret Service which, reportedly, was impressed with this service.

As information concerning the incident was made public, BIS published the news and updates on the city's Web site as well as on the city's cable station, CityTalk. The computer-aided dispatch (CAD) system moved responders quickly to the appropriate areas, while the automatic vehicle location identifier aided in positioning and tracking the surge of vehicles and personnel around the incident site. This was important because, as the event unfolded, these assets sometimes had to be redistributed along both banks of the river. Data from the BIS made this task more efficient. Emergency personnel also had up-to-the minute street and highway closure information so they could plan the quickest route to the scene, again provided through the Web-based GIS Common Operating Picture.

The demand for enhanced communications capabilities surges quickly when a major incident occurs. BIS increased cellular capacity to the incident area, increased Wi-Fi network capacity, and doubled

Internet service capacity to support the intense demand on communications reliability. Minneapolis was the first large city to use wireless Internet in support of a major disaster response effort, according to the Wireless Internet Institute. In December 2007, that organization bestowed the *Digital Cities Best Practices Award for e-Government Applications* to the City of Minneapolis in recognition of their Wi-Fi network and how it was used in broad support of emergency response after the I-35W Bridge disaster. Minneapolis won over such other finalists as Houston, Singapore, and Seoul.

The working relationships between Minneapolis and other disaster organizations were helpful in securing Federal technology assistance. FEMA provided a satellite network that assisted the JIC and PIOs to communicate with media around the world.

LOGISTICS AND PUBLIC WORKS

The EOC Logistics Section is responsible for assisting the incident Logistics Sections in assuring that all facets of the incident have the supplies and services necessary to operate. Included are base and camp facilities, food unit, security (not to be confused with law enforcement), electronic communications, medical (for rescuers), and incident transportation services.

The EOC Logistics Section Chief was a public works staff member who had extensive training in NIMS and MnNIMS. He participated in several practical exercises prior to the incident and preparedness planning for contingencies like debris management. Included within the Logistics Section were equipment, food and water, donation management, and information technology. Since the city had a long-term commitment to the incident, logistics was a dynamic process with changing needs and priorities.

The role of a public works department is sometimes easy to underestimate when disaster response is analyzed and reported. Public works can be the unsung hero whose efforts nonetheless provide critical support to all of the other functions during response and recovery. In the case of the bridge collapse, the Minneapolis Department of Public Works (DPW) played a critical and visible role. In what may be a situation unique in the country, over half of the total full-time personnel in the department, and all those with potential emergency response functions (nearly 800 employees), had received some level of NIMS training. Employees retained the lessons on Command structure and supporting roles, which was a huge help. One of the lead DPW responders noted, "I would recommend ICS training to everyone. The concept of unified command was awkward at first to deal with, but we learned firsthand [that you must] be ready anytime for anything."

Of particular note is the fact that the DPW anticipated what the needs were and took care of them. They maintained close communication with the IC and the Logistics Center. The Director of Public Works, Deputy Director, and the Traffic Engineer initially were at the EOC. They divided support into two areas: field operations and traffic management.

The field operations lead went to the Incident CP at the 10th Avenue Bridge to support the fire IC. The traffic management lead set up at 300 Border Avenue, which was the Traffic Operations Command Center. He had to identify major (hard) and local (soft) closures that were needed. They could not reach their vendor for signs, so they worked with the State's contractor.

The Bridge Maintenance Division Foreman went to the scene immediately and called in others. He had 10 employees there quickly. Division personnel put their Boston whaler in the water to help and secured barricades for traffic control to assist with the police perimeter.

DPW received many calls from vendors offering supplies and assistance. The General Foreman coordinated this and worked with the first responders on the most crucial needs, triaging all the equipment requests, determining where to deliver them, and maintaining them. Basically, DPW filled the role of quartermaster in the field. They supplied tables, chairs, fuel for generators and lights, garbage cans, portable toilets, dumpster, tarps, and generators. They also provided fencing at the Bohemian Flats through contractors, working with NTSB to make this happen.

Public works had participated in developing a sophisticated Debris Management Plan that anticipated the need to address large volumes of debris in the event a major downtown tower collapsed. With the broken bridge they had a horizontal collapse instead of a vertical structural collapse, but the volume of debris was comparable—4,000 tons of steel, and 10,000 tons of concrete. As one Minneapolis staff member noted, "We finished the plan less than a year before the collapse, so we'd thought long and hard about the issues; we weren't taken by complete surprise with no idea how to proceed."

Public works used its boat to join other rescuers at the bridge. They delivered barricades for the MPD's perimeter control and planned street closures and detours with fire and police. Since the destruction of the bridge created long-term traffic routing problems, DPW worked with the Minnesota Department of Transportation to plan solutions that would work until a new bridge is built. DPW employees installed special situation awareness cameras at key locations in and around the disaster scene. As was mentioned before, these cameras were extremely valuable in providing a common operational picture for all the Command Staff.

Figure 20. Public Works with Special Cameras.

DPW encountered a few problems, and these are being resolved before the next disaster occurs. The main roadblock was that they were not tied into the 800 MHz system that proved so useful to the other departments. The DPW was using Nextel phones and cell phones. Because of the EOC space constraints, the DPW conference room had to serve as a divisional operations center, so they were physically disconnected from the EOC. On occasion, they did not know what was happening in terms of the big picture. The Property Shop Manager obtained 20 to 30 hand-held radios for the DPW operations center and the Command Post; users were unfamiliar with how the radios operated in the 800 MHz system.

9-1-1 CENTER

The Minneapolis 9-1-1 Center was an integral part of the incident, serving as the city's primary service answering point and public safety dispatch. In 2006, the DHS recognized Minneapolis as one of the top six cities in the Nation on the DHS Tactical Interoperable Communications "scorecoard." The bridge incident was the first big test for the 800 MHz system. It became apparent early on that the 800 MHz system was a major asset. There were no incidents of fail-soft (gradual reduction of radio capacity) and only one 12-second "busy time." The 9-1-1 Center Director commented that the field providers were able to use the newer radio equipment properly. The 9-1-1 Center and the 800 MHz system performed admirably during the I-35W Bridge incident. Emergency personnel gained confidence in the new communications system, which should be good for future incidents.

The 9-1-1 Center received simultaneous notification of the collapse from citizens, the State Police, and video cameras placed on the bridge to assist with traffic status. After confirming the incident, fire dispatch worked with the fire IC to upgrade the incident. In addition, Nextel and Sprint were asked to bring their cellular on wheels (COW) vehicles to the scene to augment cell phone use.

The 9-1-1 Center was responsible for alerting key staff of the need to open the EOC. Most of the key personnel were contacted quickly except for the communications director. By 6:20 p.m., the EOC was opened and key staff members began to arrive. Additional personnel who were attending a meeting assisted with opening the EOC.

The 9-1-1 Center received over 300 calls per hour up until midnight on August 2nd. Between 9-1-1 and 3-1-1, call surge was handled with only minor inconveniences. Several challenges, however, were identified, including:

- The 9-1-1 Center needed more computers.

- Messages between the 9-1-1 Center and the EOC were handled by runners, which was acceptable. In the future, a liaison position will be considered.

- Emergency call lists should be kept current to ensure immediate contact with essential personnel.

- 9-1-1 Center operators should have greater access to critical incident stress debriefing (CISD) services.

While the Minneapolis 9-1-1 Center did an excellent job of dispatching resources, some important, generally accepted, rules should be reiterated:

- The IC is the only official who should request additional equipment, personnel, or services.

- Mutual aid or private services should not respond unless requested by the jurisdiction having authority.

- Mutual-aid units that are instructed to stage or cancel should comply immediately.

3-1-1 Operations–The Minneapolis 9-1-1 Center also runs a 3-1-1 Center that is used for nonemergency public safety and service needs. Citizens can dial 3-1-1 directly, or be transferred from 9-1-1. Originally, 3-1-1 was thought to be a tool to assist with routine needs. The bridge incident brought out 3-1-1's other strength, relieving the 9-1-1 operators from handling nonemergency calls and assisting citizens with information and directions during disasters. The 3-1-1 Center had up to 15 operators on duty, and eventually was able to screen requests from the public.

"Freelancing"–During most large multicasualty incidents, there is a tendency for mutual-aid fire and EMS agencies to self-dispatch or "freelance." Well-intentioned responders believe that their services must be needed and begin to respond prior to being requested. Another phenomenon is the misinterpretation of radio communications that may sound like an "all call" or a panic call for help.

The bridge incident was no different from others, because some units from adjacent communities responded to the incident without being requested. The 9-1-1 Center did report that, once units were instructed to cancel or stage, they complied with the order. There were accusations of "freelancing." However, by far most mutual-aid resources waited for instructions as they had learned during their NIMS training. There is evidence that an "all call" was requested from the bridge scene. According to the 9-1-1 log, at 1821 hours a unit at the bridge asked dispatch to send "all available rigs." A computer file also was found that Ramsey County Dispatch advised all available units to respond to the bridge. What is still unknown is whether Ramsey County was called, or the on-scene communication was overheard by a dispatcher.

FINANCE AND ADMINISTRATION

The Finance Department took charge of tracking emergency operations expenses. This is an important part of EOC operations, because good records enable communities to submit requests for eligible State and Federal reimbursements. It also is helpful to capture a picture of all the labor hours and overtime invested in handling a major incident, the volunteer hours, the amount and cost of supplies, acquisition costs for special equipment, food, the cost of injuries, and a host of other expenses.

The city budget officer and other staff members rotated staffing of the EOC. Purchasing, accountability, and financial risk management were coordinated by this section. Personnel immediately began to determine the operating cost of the incident by documenting as much as possible. They organized the financial records so the incident expenses could be tracked per the records management requirements of the Federal government and other sources of reimbursement, and they immediately began submissions for reimbursement.

The Minneapolis Finance Department maintained spreadsheets by department and distributed them to the heads of departments to review and confirm. Finance Department employees also tracked donations and offers of corporate assistance. It is vitally important to acknowledge publicly who donated what, so that the private sector partners receive due recognition for their role in supporting response and recovery. That, in turn, strengthens future private sector cooperation and involvement in disaster planning.

**Figure 21. Outback Steakhouse Workers
Offloading Food Donations.**

Finally, the Finance Department helped to coordinate donations and other aid with the needs communicated by the first responders (Figure 21). It is important for the EOC to have a donation management plan, and that cities anticipate and track donations and other offers of assistance. In Minneapolis, these donations involved everything from the mundane to satellite images of the bridge collapse.

MINNESOTA SECURITY AND EMERGENCY MANAGEMENT

The Minnesota Governor's Executive Order 07-14 states that "The Division of Homeland Security and Emergency Management (HSEM) shall have the coordinating role in a multiple state agency response to a disaster or emergency."[7] The I-35W Bridge collapse was an incident of the magnitude that required HSEM to open the EOC. Although there was some difficulty confirming the full extent of what had happened, HSEM opened the EOC at 6:20 p.m. The State was in immediate contact with the Minneapolis EOC to determine the city's resource status and needs.

In the Minneapolis area, the State has a role in operations, especially when hazardous materials are involved. Minnesota's Chemical Assessment Team (CAT) was deployed to the bridge to assist with hazardous materials identification, mitigation, and control. The State CAT had significant involvement in identifying styrene as the substance that had leaked out of a railcar damaged by the bridge collapse. Other State agency personnel responded to the scene and set up a Staging Area for State equipment. Prior to taking any action, HSEM personnel received approval from the fire IC. An HSEM official reported to the Minneapolis EOC as a liaison to the State.

Along with their operational duties, the HSEM EOC coordinated several continuity of operations efforts, including assisting the Governor with requesting a Federal major disaster declaration. HSEM also assisted Minneapolis with the request that the State declare the incident as a major disaster. A Federal disaster declaration was delayed because there was some uncertainty over whether or not the incident qualified for a declaration. Eventually, a partial Federal emergency declaration was confirmed. That type of declaration makes fewer Federal benefits available to assist in the recovery of business and infrastructure than a full declaration. HSEM was required to submit a separate request for supplemental SBA approval for emergency grants and access to low-interest loans for affected businesses. The city's Business Information Bureau worked with HSEM to acquire the SBA support.

After ensuring that local needs were met, HSEM focused its attention on infrastructure assessment and integrity. Although the bridge collapse caused delays in traffic movement, that complication paled when compared to the impact on the waterways. Bridge debris, hazardous materials, continuing searches for bodies, and fuel contamination from vehicles that fell into the water, required the U.S. Coast Guard to close the Mississippi River to maritime traffic between mile markers 852.7 and 853.7. Xcel Energy also reported that one of the three main electric feeder lines for the northeast portion of Minneapolis was damaged. There was fear that due to blockage of barge traffic, the Riverside Powerplant would not be able to receive coal shipments. Fortunately, the plant receives coal via rail.

By August 3rd, HSEM's recovery role became increasingly important because two barges that transport iron on the Mississippi River were unable to complete shipments. Failure to mitigate the situation quickly could have led to mass layoffs of workers whose jobs relied on navigable water. On August 6th, the city extended the emergency declaration by 30 days. The HSEM continued to coordinate Federal and State relief efforts to minimize impact on businesses. They also worked with the

---

[7] Pawlenty, T. (2007). *Executive Order 07-14 Assigning Emergency Responsibilities to State Agencies; Rescinding Executive Order 04-04.*

Minnesota Department of Transportation to open a 50-foot channel on the southern end of the collapse to restore river navigation

On August 10th, HSEM placed the State EOC back under routine operations. They were continuing to work with the DHS to justify the SBA Economic Impact Declaration Request. They also continued to support the Family Assistance Center (FAC) set up by the city and the Red Cross.

Many of HSEM's activities were behind the scenes and may not have been noticed by the average citizen. HSEM's rapid assessment of infrastructure damage and continued communications among Federal, State, and local agencies were critical in restoring business activity to the Mississippi River and its businesses.

CHAPTER VI. HAZMAT AND ENVIRONMENTAL MONITORING

The presence of hazardous materials at the scene was another concern that had to be evaluated and handled. There were numerous threats. Cars and trucks containing fuel were in the river. There was the risk of disease from decaying human remains. Much of the bridge surface and structure, with related contaminants, were submerged. Debris removal operations would create dust that could affect nearby neighborhoods. Again, numerous agencies at all levels of government were involved actively in addressing the known and potential hazardous materials, from identifying substances to monitoring the water, soil, and air. The local and State emergency management agencies, working from their EOCs, handled most of the coordination.

The MNDOT hired a contractor to check the concrete coating of the debris for asbestos levels. They reviewed as-built drawings and analyzed 38 samples taken from the collapsed structure. MNDOT concluded the bridge did not contain asbestos. Minneapolis Public Works requested a hydrologist to see if the runoff tunnels were affected. The Minnesota Pollution Control Agency (MPCA) and city staff were in charge of monitoring the shoreline for evidence of fuels at the surface. Water sampling was coordinated by the Mississippi Watershed Management Organizations in conjunction with local, State, and Federal agencies. Regular samples were taken and tested daily at both upstream and downstream locations. Tests were conducted for metals, oils, grease, polychlorinated biphenyls (PCBs), suspended solids, and other elements.

As a precaution, an air sampling plan for a half dozen sites was drawn up by MCPA working in conjunction with the U.S. Department of Environmental Protection. The goal was to determine the level of particulate matter and to monitor the air during demotion and debris removal. Air samples also were taken on Friday, August 3rd near and around the bridge and tested for lead, asbestos, and silicates. No hazardous particulates were found.

Figure 22. Railroad Car with the Leaked Substance.

A northern portion of the bridge collapsed on several railcars that were at a rail storage yard operated by Minnesota Commercial Railroad. Several piles of a white substance had leaked from one of the railcars (Figure 22). The Minnesota Department of Public Safety's Division of Homeland Security and Emergency Management (HSEM), along with MNDOT, the railroad, and MPCA ascertained that the contents of that car were nonhazardous plastic (styrene-allyl alcohol copolymer). It was fortunate that one of the other nearby railcars was not damaged by the bridge because that one contained chlorine.

The Federal Occupational Health and Safety Administration (OSHA) sent nine representatives from its scene and compliance teams who arrived at 5 p.m. on August 2nd. Concern over the possibility that pipelines may have been affected by the weight and impact of the bridge collapse were addressed by Minnesota's Office of Pipeline Safety which confirmed that there were pipelines in the vicinity of the I-35W Bridge, but that they were not affected by the incident.

CHAPTER VII. LESSONS LEARNED AND BEST PRACTICES

The I-35W Bridge collapse presented one of the most complex challenges that a city can face in an unexpected incident. Victims were located on both sides of the river, meaning there were actually two sites in terms of access and managing a response. Rescuers had to operate on dangerous and hard-to-access surfaces with the potential threat of further collapse. There were vehicles on fire. The cause of the bridge's unexpected fracture and crash into the river was uncertain at first, and there were real fears that either domestic or foreign terrorists might be involved.

The citizens of Minneapolis and government officials gave the rescue and recovery effort high marks and praised the emergency services community for an admirable performance. Why those accolades are deserved is outlined in this chapter. At the same time, our after-action analysis identified a few problem areas. The purpose of an after-action analysis is to recognize not only the strengths of a response, but to identify areas in which improvements can be made.

It is important to note that most of the concerns identified in this chapter have been recognized by the City of Minneapolis through departmental hotwash processes conducted shortly after the event. Corrective actions have been addressed by the relevant agencies and jurisdictions, but the problems are nevertheless mentioned for the sake of lessons learned that can be shared with others.

PROBLEM AREAS AND LESSONS LEARNED

1. Fire and police did not initially operate from a Unified Command Post, but managed their operations from separate Command Posts (CPs). The fire department positioned early operations from the 10th Avenue Bridge, which was the best location from which to observe and monitor rescue and fire suppression operations. Fire and police personnel maintained close radio communications but were not positioned for Unified Command until the rescues were completed. Fire Command then relocated to the police Command location. It would have been helpful if police Command had been established with fire Command on the bridge at the beginning of response activities, so that a true Unified Command operation could be managed. For example, since there were multiple CPs, the emergency medical services (EMS) Branch did not have a single point of contact.

2. Although Command personnel showed the proper regard for safety, no formal Safety Officer or Assistant Safety Officers were appointed. Also, many police and EMS personnel did not follow the Incident Commander's (IC's) orders to evacuate the bridge. Communities should ensure that all first-responder personnel are cognizant of warning and evacuation signals and understand the importance of adhering to such orders when they are communicated.

3. The EMS Branch Director did not have enough staff support to enact the EMS (Multicasualty) Branch fully. Critical incident planning should cover how to access EMS resources quickly sufficient to handle a mass casualty incident.

4. Multiple transportation groups were established, making the distribution of patients harder to track. Generally, the best procedure is to assign only one transportation group so that the status of receiving hospitals, transport resources, and patients can be managed centrally. In this inci-

dent the transportation division was divided into north and south sections with corresponding north staging and south staging areas. The breakdown in managing and tracking patients came from crews not following the established process for MCIs spelled out in the Incident Response Plan (IRP). The IRP stipulates that the Medical Resource Control Center (MRCC) will provide coordination and tracking. Only 20% of the patients were tracked by the MRCC despite repeated instructions from dispatch and EMS Branch Director.

5. Triage ribbons and tags were not used universally, which compromised the identification of patients and hospital destinations. Injured victims from the collapse were treated and transported via different modes, and the standard triage identification system was not observed. Though this caused confusion, all transported patients survived and received good medical care. Some triage tags were used by crews in the Northeast Division, but they were not used throughout the incident.

6. Well-intentioned rescuers need to keep their own safety in mind. In one case, structural turn-out gear was worn by a fire department rescuer who was operating at the edge of the water. Whenever possible, rescuers near or in the water should wear a personal flotation device.

7. An EMS official should have been part of the Emergency Operations Center (EOC). In communities where EMS is handled by the fire department, EMS Command would have been represented by senior fire officers at the EOC. Where EMS services are managed by a third party, it is essential that they participate at the EOC.

8. The EOC was of insufficient size to manage a major event. Jurisdictions should plan for backup EOC facilities that can be used if the primary EOC is inaccessible or too full.

9. The 9-1-1 EOC notification and callup did not occur as it should have because the center was overwhelmed with calls for assistance, and call lists were not updated.

10. The West Metro Medical Resource Control Center (WMMRCC) was not able to complete its mission, primarily because EMS personnel used cell phones to contact hospitals directly. EMS personnel operating at MCIs tend to perform as they do in everyday situations. Hospitals used the MNTrac System but the MCI Surge Information was not accurate because EMS personnel did not contact WMMRCC with hospital destinations and patient information.

11. For the first several hours there was confusion about which agency was the lead agency and what the roles were at the Family Assistance Center (FAC). There were problems in administration and in communication with the EOC, though services to the families were not compromised. Even at the Federal level, organizational guidelines for FACs are lacking. It is advisable for all levels of government to address how FACs should be managed and the criteria to be considered in locating, staffing, and managing this resource, including the use of volunteers.

BEST PRACTICES: NOTABLE SUCCESSES IN THE RESPONSE TO BRIDGE COLLAPSE

1. **Cooperation among first responders, mutual-aid resources, and State and Federal partners was outstanding.** Strong working relationships and knowledge of roles and procedures were arguably the greatest strengths of the Minneapolis emergency services community's response. The city had invested heavily in the development of those relationships, which were built through plan development, universal National Incident Management System (NIMS) training, appropriate use of exercises, and strategic planning over several years. These factors contributed heavily to creat-

ing an environment in which key players not only knew each other, but were familiar with the operations and disaster assignments of others. When it came time to pull together efficiently as a team–they did.

One example of how relationships made a difference can be found in the request that the Governor and the Mayor speak with one voice from the EOC to avoid the potential for releasing different information. State emergency management personnel knew their counterparts at the city's emergency management office, and a simple phone call accomplished the joint information release.

2. **Local leaders together had taken the Federal Emergency Management Agency's (FEMA's), Integrated Emergency Management course, and credited it as a major factor in their level of preparedness.** One leader who was in the city's EOC during rescue and recovery noted that, "Mt. Weather [site of the training] was a huge deal for us–it not only [helped us] identify gaps, it clarified roles and responsibilities. It was elected officials and staff from the City and County coming together to deal with some very difficult scenarios…. It's difficult to overestimate how important this was for us in charting our direction this decade."

3. **The new 800 MHz radio system streamlined communications and enabled successful connections among a variety of organizations and agencies.** The City of Minneapolis made a substantial downpayment on public protection when it purchased the new radios and system. Emergency responders stated that the system "saved our lives," was "fantastic" and "incredible."

4. **Technology played a major role in managing the response and recovery efforts.** From the real-time situational awareness provided by site video cameras, to the Web-based Geographic Information System (GIS) Common Operating Picture and traffic management systems, to the use of municipal Wi-Fi, technology was tried and tested in the I-35W Bridge incident. The technology not only performed well, it transformed the way response and recovery were handled.

5. **Responders arrived rapidly.** First-responder units arrived on scene quickly and established their respective Incident Commands based on Incident Command System (ICS) principles that affix responsibility for specific functions to subordinate Command officers.

6. **Bridge stability was evaluated quickly.** Engineers and bridge contractor personnel moved rapidly to size up the stability of the bridge and to evaluate the chances of additional collapse. When instability was suspected, fire Command made an appropriate decision to evacuate responders from the bridge until the area could be deemed safe enough for continued operations.

7. **All but one of the patients who survived the bridge collapse and were transported to a hospital, survived to discharge from the hospital.** The one patient who was pronounced dead at the hospital experienced cardiopulmonary arrest on the scene and had little to no chance of surviving.

8. **Local EMS plans worked well.** The Metro EMS Incident Response Plan (IRP) was used successfully by all EMS personnel who responded as mutual aid providers. (Mutual aid agreements are not managed through the IRP.)

9. **EMS response was rapid and in sufficient quantity.** Units approached the scene from different directions allowing for a rapid, 360-degree scene assessment.

10. **The EMS Branch used ICS.** The EMS Branch was established quickly with Divisions and Groups established. Nonambulance vehicles were used to transport patients with minor injuries to area hospitals. Non-ambulances were used only because normal access routes were blocked and alternative routing for the assigned ambulances were significantly delayed. Many of these patients had serious injuries.

11. **Hennepin County dive operations were very well organized and managed.** The Hennepin County Sheriff's Office (HCSO) established the waterside Command site shortly after the bridge failed. A dive master was put in charge of all water operations. Divers from the metro area, the Federal Bureau of Investigation (FBI), and the U.S. Navy functioned through that Command structure.

12. **Federal and military organizations brought essential resources and expertise in a timely manner to support local efforts.** Certain tasks required help from Federal and military resources. The U.S. Coast Guard, the U.S. Army Corps of Engineers, the FBI, the National Weather Service, the National Oceanic and Atmospheric Administration (NOAA), the Office of the President of the United States, and the U.S. Navy generated equipment, specially trained personnel, up-to-the-minute weather updates, and river containment engineering which strengthened local capabilities to recover the deceased and clear debris from the river.

13. **EMS dispatch had surge capacity.** Hennepin County EMS Dispatch was able to add dispatchers quickly to manage communications for the incident and for continuation of daily operations.

14. **The identities of the deceased were closely guarded until death notifications were completed.** The Minneapolis police and the Medical Examiner's staff were resolute in ensuring that families were the first to learn that their loved ones had died. Despite pressure from the press and a few other government offices, sensitive information was secured.

15. **Families of the missing and the dead were well provided for and supported during the disaster.** The FAC provided a safe place for families to gather and monitor information about the status of their relatives. Security was well provided by the Minneapolis Police Department (MPD), the Salvation Army organized meals, the Red Cross secured space and helped with donations, and the city's Department of Health and Family Support covered many of the mental health and social service requirements.

16. **The Minneapolis EOC, despite issues with its size, was able to operate in a safe and effective manner.** The EOC worked on a schedule that assured the presence of qualified professionals to manage their specialty areas. All EOC personnel were NIMS-trained and had participated in practical exercises. Early activation of a Multiagency Coordinating (MAC) Group prevented the city from being overwhelmed and enabled rapid access to community resources. The 9 p.m. Mayor's address to the community on the night of the disaster was very well done and reassured residents that response had been rapid and comprehensive. The public address set the stage for successful dissemination of information during the remainder of response and recovery. All political and administrative leaders were NIMS-trained, helping them to understand their role within the EOC and at the incident scene, and to act accordingly.

17. **There was depth among the emergency management staff.** The city communications director was out of the area when the incident occurred, so until she could return and arrive at the EOC, the deputy emergency manager, who also had experience in handling public information, was

able to temporarily fill the Public Information Officer (PIO) role. This is an excellent example of redundancy and depth in the emergency management staff.

18. **The EOC benefited from real-time images.** The strategic placement of video cameras on the bridge allowed EOC staff to view the incident scene and surrounding area in real time.

19. Special arrangements were made to provide tours of the affected area, first to family members, and then to city, county, and State elected officials, as well as the President. Enabling these leaders to view the disaster scene first hand helped create the support needed for recovery.

As a final note, it is incumbent of any report on the I-35W Bridge disaster to acknowledge the citizen heroes who courageously helped to rescue and provided aid and comfort to survivors during those first awful minutes and hours. Their selflessness is a tribute to them and to their community.